DICTIONARY OF
BIOINFORMATICS

DICTIONARY OF BIOINFORMATICS

K. Manikandakumar

Lecturer
Department of Physics
Bharathidasan University College (W)
Orathanadu, Tanjore District
Tamil Nadu

Chennai   Trichy   Tirunelveli   NewDelhi

ISBN 978-81-8094-077-4   **MJP Publishers**

All rights reserved   No. 44, Nallathambi Street,
Triplicane, Chennai 600 005

MJP 064   © Publishers, 2019

Publisher : C. Janarthanan

To

My
beloved mother,
brothers and sisters,
family members, teachers
and
the Almighty God

PREFACE

Bioinformatics is a field that presents one of the grand challenges of today. The field is faced with a strong demand for immediate solutions, because the genomic and post-genomic data that are being collected harbour many biological insights whose deciphering can be the basis for dramatic scientific and economical success, and is promising to have a large impact on society.

Recent developments in molecular biology, genomics, proteomics, and other areas have produced a wealth of experimental data on sequences and three-dimensional structures of biological macromolecules. Consequently, the service of various computational methods of analysis to obtain useful information has now turned out to be a major new discipline as Bioinformatics.

The *Dictionary of Bioinformatics* is a vital reference for important terms, offering students and researchers a convenient summary of the core knowledge of the field, with concise and accurate definitions of over 2200 words, phrases, and concepts in this book.

Anyone working in basic sciences and life sciences field of research today will find the *Dictionary of Bioinformatics* to be an essential companion.

I invite constructive comments and valued suggestions from the students and learned teachers and researchers for further improvement of the book in subsequent editions. Any errors or omissions are wholly inadvertent and will be corrected at the first opportunity.

I am grateful to the many people including friends and colleagues who have supported me in many ways. I extend my thanks to the following eminent people:

Dr. M. Ponnavaiko, Vice Chancellor, Bharathidasan University

Dr. T. Ramaswamy, Registrar, Bharathidasan University

Dr. C. Thangamuthu, former Vice Chancellor, Bharathidasan University

Dr. V. Radhakrishnan, former Registrar, Bharathidasan University

Dr. S. Parthasarathy, Reader in Bioinformatics, Bharathidasan University.

I express my heartfelt thanks to Mr. J.C. Pillai, Mr. Sajeesh Kumar and the editorial team of MJP Publishers, Chennai, for their interest shown in preparing the book and for the superb production.

Finally, I would like to express my deep gratitude to my family members for bearing with my physical or mental absence while preparing this book. They gave the most for receiving the least.

K. Manikandakumar

CONTENTS

DICTIONARY OF
BIOINFORMATICS

DICTIONARY OF
BIOINFORMATICS

K. Manikandakumar

Lecturer
Department of Physics
Bharathidasan University College (W)
Orathanadu, Tanjore District
Tamil Nadu

ISBN 978-81-8094-077-4   **MJP Publishers**

All rights reserved   No. 44, Nallathambi Street,
Printed and bound in India   Triplicane, Chennai 600 005

MJP 064   © Publishers, 2016

Publisher : C. Janarthanan

To

My
beloved mother,
brothers and sisters,
family members, teachers
and
the Almighty God

PREFACE

Bioinformatics is a field that presents one of the grand challenges of today. The field is faced with a strong demand for immediate solutions, because the genomic and post-genomic data that are being collected harbour many biological insights whose deciphering can be the basis for dramatic scientific and economical success, and is promising to have a large impact on society.

Recent developments in molecular biology, genomics, proteomics, and other areas have produced a wealth of experimental data on sequences and three-dimensional structures of biological macromolecules. Consequently, the service of various computational methods of analysis to obtain useful information has now turned out to be a major new discipline as Bioinformatics.

The *Dictionary of Bioinformatics* is a vital reference for important terms, offering students and researchers a convenient summary of the core knowledge of the field, with concise and accurate definitions of over 2200 words, phrases, and concepts in this book.

Anyone working in basic sciences and life sciences field of research today will find the *Dictionary of Bioinformatics* to be an essential companion.

I invite constructive comments and valued suggestions from the students and learned teachers and researchers for further improvement of the book in subsequent editions. Any errors or omissions are wholly inadvertent and will be corrected at the first opportunity.

I am grateful to the many people including friends and colleagues who have supported me in many ways. I extend my thanks to the following eminent people:

Dr. M. Ponnavaiko, Vice Chancellor, Bharathidasan University

Dr. T. Ramaswamy, Registrar, Bharathidasan University

Dr. C. Thangamuthu, former Vice Chancellor, Bharathidasan University

Dr. V. Radhakrishnan, former Registrar, Bharathidasan University

Dr. S. Parthasarathy, Reader in Bioinformatics, Bharathidasan University.

I express my heartfelt thanks to Mr. J.C. Pillai, Mr. Sajeesh Kumar and the editorial team of MJP Publishers, Chennai, for their interest shown in preparing the book and for the superb production.

Finally, I would like to express my deep gratitude to my family members for bearing with my physical or mental absence while preparing this book. They gave the most for receiving the least.

K. Manikandakumar

CONTENTS

A

abduction A method of reasoning in which one chooses the hypothesis that would, if true, best explain the relevant evidence. It is also used to just mean the generation of hypotheses to explain observations or conclusions.

ab initio A method that derives a 3D structure from initial physical forces and interactions.

abiogenesis An early theory that held some organisms originated from non-living material.

acceptor arm The part of a tRNA molecule to which an amino acid attaches during translation.

acceptor site The splice site at the 3′ end of an intron.

accessible surface The surface that is traced by the centre of a probe molecule (usually water) as it rolls on the van der Waals surface of a molecule.

accession number An identifier supplied by the curators of the major biological databases upon submission of a novel entry that uniquely identifies that sequence (or other) entry.

AceDB A genome database system originally developed for the *C. elegans* genome project, from which its name was derived (A *C. elegans* DataBase). The tools in it have been generalized and the same software is now used for many different genomic databases.

acentric fragment A piece of chromosome without a centromere.

acidic domain A type of activation domain.

acridine dye A chemical compound that intercalates between adjacent base pairs of the double helix, which causes a frameshift mutation.

acrocentric A chromosome in which the centromere lies very near to one end.

activation The time-varying value that is the output of a neuron.

activation domain The part of a transcription factor that makes contact with the initiation complex.

activation function A function that translates a neuron's net input to an activation value.

activator A DNA-binding protein that increases the rate of transcription initiation at a promoter.

active site A specific region of an enzyme where the substrate binds. It is the amino acid residues present at the catalytic site of an enzyme. These residues provide the binding and activation energy needed to place the substrate into its transition state and bridge the energy barrier of the reaction undergoing catalysis.

acyclic graph A directed graph with no directed cycles, i.e., for any vertex v, there is no non-empty directed path that starts and ends on v. It is also called directed acyclic graphs (DAG). DAGs appear in models where it doesn't make sense for a vertex to have a path to itself, for example, it an edge $u \rightarrow v$ indicates that v is a part of u, such a path would indicate that u is a part of itself, which is impossible. A DAG flows in a single direction. Each directed acyclic graph gives rise to a partial order $\leq$ on its vertices, where $u \leq v$ exactly when there exists a directed path for u to v in the DAG. However, many different DAGs may give rise to this same reachability relation. Among all such DAGs, the one with the fewest edges is the transitive reduction of each of them and the one with the most is their transitive closure. In particular, the transitive closure is the reachability order $\leq$.

acylation The attachment of a lipid side chain to a polypeptide molecule.

Ada enzyme An enzyme in *Escherichia coli* that is involved in the direct repair of alkylation mutations.

adaptability The potential or ability of a population to adapt to changes in the environmental condition through changes of its genetic structure.

adaptation An internal change in a system in response to an external event in the system's environment.

adaptive Subject to adaptation, i.e., able to change over time to improve fitness.

adaptive radiation Evolutionary diversification of a generalized ancestral form with the production of a number of specialized forms by adaptation.

adaptive surface A surface plotted in three dimensions with all possible combinations of allele frequencies for different loci plotted along the plane, and the mean fitness for each combination plotted as the height of the surface.

additive genetic variance The proportion of genetic variation that is the summation of the effect of all individual genes influencing a trait. It is the average effect of substituting one allele for another.

additive tree A phylogenetic tree in which the distance between any two terminal nodes is equal to the sum of the branch length connecting them.

additivity A metric space is additive if and only it satisfies the "four point condition".

adenine (A) A purine (nitrogenous base) found in DNA and RNA that form base pairs with thymine (A-T) and uracil (A-U).

adenosine The nucleoside containing adenine as its base.

Adenosine Deaminase (ADA) Deficiency A severe immuno-deficiency disease that results from a lack of an enzyme adenosine deaminase. It usually leads to death within the first few months of life.

Adenosine Diphosphate (ADP) Lower energy form of ATP, having two (instead of the three in ATP) phosphate groups attached to the adenine base and ribose sugar.

Adenosine Triphosphate (ATP) A common form in which energy is stored in living systems. It consists of a nucleotide (with ribose sugar) with three phosphate groups.

adenylate cyclase An enzyme that converts ATP to cyclic AMP.

A-DNA One of the many possible double-helical structures of DNA. It is a right-handed double helix, fairly similar to the more common and well-known B-DNA form, but with a shorter, more compact, helical strucure. A-DNA is thought to be one of the three biologically active double-helical structures along with B- and Z-DNA. It appears likely that it occurs only in dehyrated samples of DNA, such as those used in crystallographic experiments, and possibly is also assumed by DNA–RNA hybrid helices and by regions of double-stranded RNA.

adjacent segregation Segregation of a heterozygous reciprocal translocation in which a translocated chromosome and a normal chromosome segregate together, producing an aneuploid gamete.

adjacent-1 segregation Segregation from a heterozygous reciprocal translocation in which homologous centromeres go to oppositie poles of the first-division spindle.

adjacent-2 segregation Segregation from a heterozygous reciprocal translocation in which homologous centromeres to the same pole of the first-division spindle.

advantageous mutation A mutation that increases the fitness of the organism carrying it.

affine An equation that can be written in terms of matrix–vector multiplication and vector addition.

affine gap cost A scoring system for gaps within alignments that charge a penalty for the existence of a gap and an additional per-residue penalty proportional to the gap's length.

affinity chromatography A column chromatography method that uses a ligand that binds to the molecule being purified.

agarose gel electrophoresis Electrophoresis carried out in an agarose gel, used to separate DNA molecules whose length is between 100 and 50,000 bp.

agent Computer software constructed to operate with a degree of autonomy from its user. It is a independent, autonomous, software module that can search the Internet for data or content pertinent to a particular application, such as a gene, protein, or biological system.

agonist A molecule that produces the same or elevated effect as a natural substrate or effector molecule.

agricultural biotechnology The application of recombinant DNA (rDNA) technology to plants and organisms commonly used in agriculture.

Alanine (Ala, A) One of the 20 amino acids and its codons are GCA, GCC or GCU. In mammals, it is a non-essential amino acid with a molecular formula $CH_3 - CH(NH_2) -COOH$.

AlChemy A computer system for simulating polymer reactions.

algorithm A detailed, unambiguous sequence of instructions that describe how to perform a particular computational task. A fixed procedure embodied in a computer program. It is a series of steps defining a procedure or formula for solving a problem, that can be coded into a programming language and executed. Bioinformatics algorithms typically are used to process, store, analyse, visualize and make predictions from biological data.

algorithmic complexity The size of the smallest program that can produce a particular sequence of numbers.

alias An abbreviation for a frequently used command or series of commands.

alignment 1. The process of lining up two or more sequences to achieve maximal levels of identity (conservation, in the case of amino acid sequences) for the purpose of assessing the degree of similarity and the possibility of homology. 2. The result of a comparison of two or more gene or protein sequences in order to determine their degree of base or amino acid similarity. Sequence alignments are used to determine the similarity, homology, function or other degree of relatedness between two or more genes or gene products.

alignment score A numerical value that describes the overall quality of an alignment. Higher numbers correspond to higher similarity.

alkylating agent A mutagen that acts by adding alkyl groups to nucleotide bases.

all-beta A class that includes two major fold groups: sandwiches and barrels.

allele Variant form of a gene that occupies a specific position or locus on a chromosome.

allele frequency The frequency of an allele in a population.

Allele-Specific Oligonucleotide Hybridization (ASO hybridization) The use of an oligonucleotide probe to determine which of two alternative nucleotide sequences is contained in a DNA molecule.

allogeneic Variation in alleles among members of the same species.

allopolyploid A polyploid nucleus derived from the fusion of gametes from different species.

alphabet The set of all possible symbols in an application.

alpha helix A common secondary structure formed by segments of polypeptides.

alphoid DNA The tandemly repeated nucleotide sequences located in the centromeric regions of human chromosomes.

alternative splice-form One of the possible alternate

combinations of exons into a folded protein that are possible by recombining multiple gene segments during mRNA splicing in higher organisms.

alternative splicing The production of two or more mRNA molecules from a single pre-mRNA sequence by using different acceptor and donor sites.

Alu A short, repetitive string of about 300 DNA bases that are scattered throughout the genome, often in regions that have many genes.

***Alu* family** A common set of dispersed DNA sequences found throughout the human genome; each is about 300 bases long and they are repeated at least 500,000 times. *Alu* sequences are speculated to have originated from viral RNA sequences that integrated into human DNA, thousands of years ago.

***Alu* PCR** A clone fingerprinting technique that uses PCR to detect the relative positions of *Alu* sequences in the cloned DNA fragments.

AMBER A type of molecular mechanics force field used to predict the 3D structure of a protein.

amber mutation Mutation of the UAG termination codon.

American Medical Informatics Association (AMIA) An organization dedicated to developing and using information technologies to improve health care.

American National Standards Institute (ANSI) A private, non-profit organization that administers and coordinates the U.S. voluntary standardization and conformity assessment system.

amino acid The basic building block of proteins. Chemical building blocks that are joined by amide (peptide) linkages to form a polypeptide chain of a protein.

amino acid alphabet A twenty-character alphabet (characters A, C, D, E, F, G, H, I, K, L, M, N, P, Q, R, S, T, V, W, Y), each representing one of the twenty amino acids coded by DNA.

aminoacylation Attachment of an amino acid to the acceptor arm of a tRNA molecule.

aminoacyl site (A-site) The site in the ribosome occupied by the aminoacyl-tRNA during translation.

aminoacyl-tRNA synthetase An enzyme that catalyses the aminoacylation of one or more tRNA molecules.

amino terminus The -NH$_2$ (amino) end of a polypeptide.

amniocentesis A procedure used in antenatal diagnosis of genetic abnormality in which a sample of the amniotic fluid surrounding the foetus is removed, and either DNA is prepared directly from the amniocytes suspended in it or these cells may be cultured and their chromosomes can be examined.

amniocyte A foetal cell suspended within the amniotic fluid surrounding the foetus.

amphiphilic Having a polar water-soluble group attached to a water-insoluble hydrocarbon chain.

amplification An increase *in vivo* or *in vitro* in the number of copies of a specific DNA fragment. It is the process of repeatedly making copies of the same piece of DNA.

anagenesis An evolutionary process in which the act of one species evolving into another without a split in the phylogenetic tree.

analog Biological structures which perform similar functions by similar mechanisms but evolved separately, e.g. a chemical so similar to one of the normal bases that it can be incorporated into DNA.

analogous structures Body parts that serve the same function in different organisms, but differ in structure and embryological development, e.g. the wings of insects and birds.

analogy Similarity by convergent evolution, but not by common evolutionary ancestry. Identifying analogous or homologous genes via similarity searching and alignment is one of the chief uses of bioinformatics. (*See* also alignment, similarity search.)

analytical Represented symbolically in a closed form that does not require any of the complex aspects of a program such as an iterative sum.

analytical solution An exact solution to a problem that can be calculated symbolically by manipulating equations.

anaphase A stage in a cell division at which the chromosomes move to opposite ends of the spindle.

ancestral character state A character state possessed by a remote common ancestor of a group of organisms.

ancient DNA DNA preserved in ancient biological material.

aneuploidy The presence of extra chromosomes, such that the chromosomal composition of a cell is not an exact multiple of the haploid set.

animal model A laboratory animal useful for medical research because it has specific characteristics that resemble a human disease or disorder.

annealing Attachment of an oligonucleotide primer to a DNA or RNA template.

annotated databases Databases may contain a combination of amino acid sequences, comments, literature references and notes on known post-translational modifications to the sequence. A database that contains all of these elements is referred to as "annotated". Other databases contain only the sequence, an accession number and a descriptive title. Annotation of each entry is obviously very time-consuming and difficult to maintain without errors. Therefore, annotated databases usually have many fewer sequence entries than non-annotated ones. Annotation also implies that some functional or structural information is known about the mature protein, as opposed to a sequence that is known only from the translation of a stretch of nucleotide sequence. Even the best annotated databases now include large numbers of entries that have very little real information about the mature protein other than some reference to

who sequenced and translated the nucleotide sequence. Annotated databases are technically superior for many purposes, because they contain information about the true form of the mature protein.

annotation A combination of comments, notations, references, and citations, either in free format or utilizing a controlled vocabulary, that together describe all the experimental and inferred information about a gene or protein. Annotations can also be applied to the description of other biological systems. Batch, automated annotation of bulk biological sequence is one of the key uses of bioinformatics tools.

antibody A blood protein that is produced in response to and counteracts an antigen.

anticipation The process whereby some genetic diseases get more severe in each successive generation.

anticodon The triplet of nucleotides, at positions 34–36 in a tRNA molecule, that form base pairs with a codon in an mRNA molecule. It is the triplet of contiguous bases on tRNA that binds to the codon sequence of nucleotides on mRNA, e.g. the codon for glycine is GGG. The anticodon for glycine is CCC.

anticodon arm A part of the structure of a tRNA molecule that includes the anticodon.

antigen Any foreign molecule that stimulates an immune response in a vertebrate organism. Many antigens are proteins such as the surface proteins of foreign organisms.

antigenic index A prediction of the antigenicity of a sequence based on its hydrophilicity, predicted side chain flexibility, surface probability, and the predicted turns.

antiparallel The opposite orientation of the two complementary strands of a DNA duplex. The opposite orientation of two beta strands to create a beta sheet.

antisense DNA or RNA composed of the complementary sequence to the target DNA/RNA. Also used to describe a therapeutic strategy that uses antisense DNA or RNA sequences to target specific gene DNA sequences or mRNA implicated in disease, in order to bind and physically inhibit their expression by physically blocking them.

antisense strand The noncoding strand in double-stranded DNA. The antisense strand serves as the template for mRNA synthesis.

antitermination The prokaryotic cell's aid to fix premature termination of RNA synthesis during the transcription of DNA. It occurs when the RNA polymerase ignores the termination signal, and it provides a mechanism whereby one or more genes at the end of an operon can be switched either on or off, depending on the polymerase either recognizing or not recognizing the termination signal.

antiterminator protein A protein that attaches to bacterial DNA and mediates antitermination.

AP endonuclease An enzyme that is involved in base excision repair.

AP site A nucleotide position in a DNA molecule where the nitrogenous base component is missing.

apomorphic character state A character state that evolved in a recent ancestor of a subset of organisms in a group being studied.

apoprotein A protein without its coenzymes, cofactors and prosthetic groups that are required for its functionality.

apoptosis Programmed cell death, the body's normal method of disposing of damaged, unwanted cells.

application A computer program that carries out a specific type of task.

Arabidopsis thaliana Commonly known as mouse-eared cress, the first plant species to have its genome sequenced and available to the public as a model for genomics research.

archaea One of the two main groups of prokaryotes, mostly found in extreme environments.

Arden syntax A language created to encode actions within a clinical protocol into a set of situation-action rules, for computer interpretation, and also to facilitate exchange between different institutions.

arginine (Arg, R) An amino acid.

arithmetic mean The sum of a set of values divided by the number of values.

arity The number of attributes (columns) in a database relation (table).

arms race A phenomenon of evolution in which two or more species adapt to one another in a co-evolutionary manner.

arrayed library Individual primary recombinant clones (hosted in phage, cosmid, YAC, or other vector) that are placed in two-dimensional arrays in microtitre dishes.

array of hairpins An assemble of alpha-helices that cannot be described as a bundle or a folded leaf.

Artificial Intelligence (AI) The study or application of techniques to emulate "intelligent" behaviours, such as planning, vision, or language processing.

Artificial Intelligence in Medicine (AIM) The application of artificial intelligence methods to solve problems in medicine such as diagnosis or therapy planning.

artificial life The study of biological processes within the confines of a computer.

Artificial Neural Network (ANN) A network of neurons that are connected graphically through synapses or weights.

ascertainment In scientific research, ascertainment bias occurs when false results are produced by non-random sampling and conclusions made about an entire group are based on a distorted or nontypical sample. Ascertainment bias may be easy to recognize or difficult to detect.

ASCII The coding system used to represent characters, where each character that can be used by the computer is described by a unique numeric code.

ascospore One of the haploid products of meiosis in an ascomycete such as brewer's yeast, *Saccharomyces cerevisiae*.

ascus The structure that contains the four ascospores produced by a single meiosis of the yeast *Saccharomyces cerevisiae*.

Asia-Pacific Bioinformatics Network (APBioNet) A non-profit, non-governmental, international organization focused on the promotion of bioinformatics in the Asia-Pacific region.

ASN.1 Abstract Syntax Notation One. An international standard that specifies data used in communication protocols.

asparagine (Asn, N) One of the 20 most common natural amino acids on earth. It has carboxamide as the side chain's functional group. It is not an essential amino acid. Its codons are AAU and AAC.

aspartic acid (Asp, D) An α-amino acid with the chemical formula $HO_2CCH(NH_2)CH_2CO_2H$. The carboxylate anion of aspartic acid is known as aspartate. The L-isomer of aspartate is one of the 20 proteinogenic amino acids, i.e., the building blocks of proteins. Its codons are GAU and GAC.

assay Analysis used to determine the presence, absence, or quantity of a substance or biological activity of the sample.

assembly The process of correctly joining together the DNA sequences from individual sequencing experiments into a contiguous segment. Compilation of overlapping sequences from one or more related genes that have been clustered together based on their degree of sequence identity or similarity. Sequence assembly may be used to piece together "shotgun" sequencing fragments (*see* also shotgun sequencing) based upon overlapping restriction enzyme digests, or may be used to identify and index novel genes from "single-pass" cDNA sequencing efforts. It is the process of placing fragments of DNA that have been sequenced into their correct position within the chromosome.

associative memory Memory that can be referenced by content, rather than by location.

assortative mating The tendency for mates to be chosen non-randomly.

assortment The movement of homologous chromosomes to opposite poles in the first meiotic division and chromatids in the second meiotic division leading to the segregation of alleles.

asymmetrical exon An exon flanked by introns of different phase classes.

asymmetric unit In a crystal, the level at which there is no symmetry. For example, the alpha–beta dimer can be considered to be the asymmetric unit of the haemoglobin tetramer in solution.

asynchronous Occurring independently, but on a similar timescale.

asynchronous communication A mode of communication between two or more parties in which the exchange does not require the parties to participate at the same time.

ataxia-telangiectasia A rare fatal disease involving a damaged immune system, unsteady walk, premature aging, and a strong predisposition to some kinds of cancer.

atomicity A principle of database design that requires all transactions (to add, modify, or delete data) to follow an "all or nothing" rule.

attenuation A process used by some bacteria to regulate expression of an amino acid biosynthetic operon in accordance with the levels of the amino acid in the cell.

attractor A characterization of the long-term behaviour of a dissipative dynamical system.

attribute A single data item related to a database object.

AU–AC intron A type of intron found in eukaryotic nuclear genes: the first two nucleotides in the intron are 5´-AU-3´ and the last two are 5´-AC-3´.

Autoimmune Lympho Proliferative Syndrome (ALPS) A human disease caused by failure of lymphocytes to die once they have finished doing their job. As a result, the spleen and lymph nodes grow large, and immune cells may attack the body's own tissues, a condition known as autoimmunity.

autonomous agent An entity with limited perception of its environment that can process information to calculate an action so as to be goal-seeking on a local scale.

Autonomously Replicating Sequence (ARS) A DNA sequence, especially from yeast, that confers replicative ability on a non-replicative plasmid.

autopolyploid A polyploid nucleus derived from the fusion of two non-haploid gametes from the same species.

autoradiography A technique that uses X-ray-sensitive photographic film to detect and visualize radioactively labelled molecules.

autosomal A position on any chromosome other than a sex determining chromosome.

autosome A chromosome that is not involved in sex determination.

auxotroph A mutant microorganism that can grow only when supplied with a nutrient that is not needed by the wild type.

awk A pattern scanning and processing language for UNIX.

axiom A statement that is assumed to be true.

B

B An amino acid that is either aspartic acid or glutamic acid.

backcross A cross between an animal that is heterozygous for alleles obtained from two parental strains and a second animal from one of those parental strains.

backpropagation An algorithm for efficiently calculating the error gradient of a neural network, which can then be used as the basis of learning.

backtracking algorithm The process of repeatedly exploring paths until you encounter the solution.

BAC library A resource that identifies a sequence of bacterial artificial chromosomes (BACs).

bacteria Structurally simple, single-celled organisms that have no nucleus.

Bacterial Artificial Chromosome (BAC) A recombinant DNA plasmid in which a large fragment of DNA has been ligated into a suitable vector, making possible its replication and segregation in bacterial cells.

bacterial bioinformatics Antibiotic resistance amongst virulent species is on the increase, causing major concern worldwide. It is evident that some current therapies are no longer effective, and accordingly, novel antimicrobials will need to be developed. To help counteract these problems, advances in technology can be used to hasten the hunt for new drug and vaccine targets. One obvious advantage of using computer-based screening techniques (bioinformatics) to scan newly sequenced pathogen genomes is the speed at which identification of novel targets can be carried out.

bacteriophage   A virus that infects bacteria.   The bacteriophage DNA has served as a basis for cloning vectors, and is also utilized to create phage libraries containing human or other genes.

baculovirus   An insect virus which forms the basis of a protein expression system.

balanced   polymorphism   A polymorphism that is stable over time and is maintained by balanced selection.

balanced selection   A selection regime that results in the maintenance of two or more alleles at a locus in a population, e.g. overdominance.

balanced tree   A common data structure for creating indexes on disk.

bandwidth   The amount of data that can be transmitted across a communication channel in a given period of time.

BankIt   A web-based tool for submitting DNA or RNA sequences to GenBank.

Barr body   The highly condensed chromatin formed by an inactivated X chromosome.

barrier to gene flow A factor, such as geographic, mechanical, and behavioural isolating mechanisms that restrict gene flow between populations, leading to populations with differing allele frequencies.

basal promoter   The position within a eukaryotic promoter where the initiation complex is assembled.

basal rate of transcription The number of productive initiations of transcription occurring per unit time at a particular promoter.

base   One of five molecules which   are   assembled,   along with a ribose and a phosphate, to   form   nucleotides.Adenine (A), guanine (G), cytosine (C), and thymine (T) are found in DNA while RNA is made from adenine (A), guanine (G), cytosine (C), and uracil (U).

base analog A chemical compound that resembles and can replace one of the nitrogenous bases normally found in DNA.

base excision repair A DNA repair process that involves excision and replacement of an abnormal base.

baseless site   A position in a DNA molecule where the base component of the nucleotide is missing.

base pair (bp)   A pair of nitrogenous bases (a purine and a pyrimidine), held together by hydrogen bonds, that form the core of DNA and RNA, i.e., the A : T, G : C and A : U interactions. The atomic structure of these bases preselect the pairing of adenine with thymine (uracil in RNA) and the pairing of guanine with cytosine.

base sequence The order of nucleotide bases in a DNA or RNA molecule.

base sequence analysis A method, sometimes automated, for determining the base sequence of DNA or RNA.

base stacking The hydrophobic interactions that occur between adjacent base pairs in a double-stranded DNA molecule.

basic domain A type of DNA-binding domain.

basin of attraction The subset of states of a dynamical system's state space that converge into an attractor.

basin portrait The set of basins of attraction for a dynamical system.

batch Entrez A feature in Entrez that allows the retrieval of many sequences at once and saves the sequences to a file on a local computer.

Baum–Welch algorithm An algorithm to find parameters A, B, and Π of a hidden Markov model that have the maximum likelihood of generating the observed sequence of output symbols.

Bayesian network (BN) A directed, acyclic graph in which nodes represent stochastic variables (either continuous variables or with discrete states) and the edges represent probabilistic influences (represented as conditional probabilities).

Bayes theorem Theorem used to calculate the relative probability of an event, given the probabilities of associated events.

B chromosome A chromosome possessed by some individuals in a population, but not all.

B-DNA The most common structural conformation of the DNA double helix in living cells, which exist as a right-handed helix.

beads-on-a-string An unpacked form of chromatin that consists of nucleosome beads on a string of DNA.

BEAUTY BLAST Enhanced Alignment Utility. A tool, developed at Baylor College of Medicine (Worley *et al.* 1995), which uses BLAST to search several custom databases and incorporates sequence family information, location of conserved domains, and information about any annotated sites or domains directly into the BLAST query results.

behavioural genetics The study of genes that may influence behaviour.

behavioural strategy A type of game-theoretic strategy.

BEST map Map indicating the correlation between bacterial artificial chromosome and expressed sequence tags.

beta function A function of two variables used in probability distributions.

beta-sheet A common secondary structure of a protein, in which several beta strand (or elongated) regions of a protein line up next to each other to form a flat sheet structure. It is a three-dimensional arrangement taken up by polypeptide chains that consists of alternating strands linked by hydrogen bonds. The alternating strands together form a sheet that is frequently twisted.

beta turn A sequence of four amino acids, the second of which is usually glycine, that causes a polypeptide to change direction.

bias A threshold value for input to a neuron, such that the neuron will not activate ("fire") for input values below that value.

bifurcation The splitting of a single mode of a system's behaviour into two new modes. It is the graphical representation in a phylogenetic tree of an evolutionary speciation event whereby an ancestral taxon splits into two.

big-O notation (O) A theoretical measure of the execution of an algorithm, usually the time or memory needed, given the problem size n, which is usually the number of items.

binary A computer code written in a form that uses only 0s and 1s. A string of bits.

binomial distribution A probability distribution for the number of times that an outcome with constant probability will occur in a succession of repetitions of a statistical experiment.

biochemical pathway A network of interacting molecules that is responsible for a specific biochemical junction, such as a metabolic pathway or a signal transduction pathway.

biocorba.org Provides an object- oriented, language neutral, platform independent method for describing and solving bioinformatic problems.

BioCORBA's mission is to leverage the code of the other Bio projects in a simple and easy to use fashion. For example, language neutral environment allows users to write programs using BioPython and access BioPerl modules through the CORBA server.

biodiversity Biological diversity that can be measured in terms of genetic, species, or ecosystem diversity.

bioinformatics The application of information technology to the field of molecular biology. The term bioinformatics was coined by Paulien Hogeweg in 1978 for the study of informatic processes in biotic systems. Bioinformatics nowadays entails the creation and advancement of databases, algorithms, computational and statistical techniques, and the theory to solve formal and practical problems arising from the management and analysis of biological data. Over the past few decades rapid developments in genomic and other molecular research technologies, and developments in information technologies have combined to produce a tremendous amount of information related to molecular biology. It is the name given to these mathematical and computing approaches used to glean understanding of biological processes. Common activities in bioinformatics include mapping and analysing DNA and protein sequences, aligning different DNA and protein sequences to compare them, creating and viewing 3D models of protein structures. Major research efforts in the field include sequence alignment, gene finding, genome assembly, protein-structure alignment, protein-structure prediction, prediction of gene expression and protein–protein interactions and the modelling of evolution.

Bioinformatic Sequence Markup Language (BSML) An extensible language specification and container for bioinformatic data. Encodes biological sequence information and includes graphical representations of biologically meaningful objects such as sequences, genes, electrophoresis gels, and multiple alignments.

biojava.org An open-source project dedicated to provide Java tools for processing biological data. This will include objects for manipulating sequences, file parsers, CORBA interoperability, access to AceDB, dynamic programming, and simple statistical routines. The BioJava library is useful for automating those daily and mundane bioinformatics tasks.

BioLisp.org A public resource supporting scientists who use Lisp to develop intelligent applications in the biological sciences.

biological databases A database that have inherent complications stemming from the nature of the information they contain and the dependence of computational methods on these data. Most biological data are not digital, making machine-readability of the data (for automated data-mining) impossible. In addition, the lack of standardized nomenclature and ontology, the use of protein aliases (leading to ambiguity), the lack of interoperability across databases, and the presence of errors in database annotations have hindered and complicated the use of computational methods.

biological information The information contained in the genome of an organism that directs the development and maintenance of that organism.

Biomedical Informatics Research Network (BIRN) An initiative sponsored by the National Institute of Health (NIH)-National Center for Research Resources (NCRR), which fosters large-scale biomedical science collaborations by utilizing emerging computing infrastructure.

bioMOBY An international group of biological data hosts, biological data service providers, and coders whose aim is to set standards for biological data representation, distribution and discovery.

BIONLP.org Natural language processing of biology text.

Bio-ontology Standards Group An effort that unde ways to standardize the domain-specific ontologies and vocabularies to support interoperability of data and software components.

BioPAX Biological Pathways Exchange. A collaborative effort to create a data exchange format for biological pathways data.

bioperl.org An international association of developers of open source Perl tools for bioinformatics, genomics and life science research. The bioperl server provides an online resource for modules, scripts, and web links for developers of Perl-based software for life science research.

biopython.org An international association of developers of freely available Python tools for computational molecular biology. biopython.org provides an online resource for modules, scripts, and web links for developers of python-based software for life science research.

bioremediation The use of biological organisms such

as plants or microbes to aid in removing hazardous substances from an area.

biotechnology Technology based on biology, especially when used in agriculture, food science, and medicine, i.e., any technological application that uses biological systems, living organisms, or derivatives thereof, to make or modify products or processes for specific use.

Biotechnology Industry Organization (BIO) An organization for bio-technology information, advocacy and business support.

BioWidget Consortium Home Page The BioWidgets tool kit is a collection of Java Beans used for the development of graphic applications and/or applets in the genomics domain.

bioxml.org A site created to be a centre for development of the open source biological DTDs.

bipartite graph A graph whose vertices can be partitioned into two disjoint subsets, U and V, such that each edge connects a vertex from U to one from V.

birth defect A structural, functional, or metabolic abnormality present at birth that results in physical or mental disability or is fatal.

BISTI Biomedical Information Science and Technology Initiative. NIH working group on biomedical computing made recommendations in their June 3, 1999 report on creating a national bioinformatics infrastructure.

BISTI Consortium Established in May 2000 to serve as the focus of biomedical computing issues at the NIH and to facilitate implementation of the BISTI recommendations. The Consortium is composed of senior-level representatives from the NIH centres and institutes, and representatives of other federal agencies concerned with bioinformatics and computational applications. The mission of the BISTI Consortium is to make optimal use of computer science and technology to address problems in biology and medicine by fostering new basic understandings, collaborations, and transdisciplinary initiatives between the computational and biomedical sciences.

bit The fundamental unit of information, a binary (base-2) digit.

bit score (S′) The value S′ is derived from the raw alignment score S, in which the statistical properties of the scoring system used have been taken into account. Because bit scores have been normalized with respect to the scoring system, they can be used to compare alignment scores from different searches.

bivalent The structure formed when a pair of homologous chromosomes lines up during meiosis.

BLAST Basic Local Alignment Search Tool. An algorithm for comparing primary biological sequence information, such as the amino acid sequences of different proteins or the nucleotides of DNA sequences. A BLAST search enables a researcher to compare a query sequence with a library or database of sequences, and identify library sequences that resemble the query sequence above a certain threshold. For example, following the discovery of a previously unknown gene in the mouse, a scientist will typically perform a BLAST search of the human genome to see if humans carry a similar gene; BLAST will identify sequences in the human genome that resemble the mouse gene based on similarity of sequence. The BLAST program was designed by Eugene Myers, Stephen Alteshul, Warren Gish, David J. Lipman and Webb Miller at the NIH and was published in *J. Mol. Biol.* in 1990.

BLAST2 A newer release of BLAST that allows for insertions or deletions in the sequences being aligned.

blending Term applied to 19th century belief that parental traits "blended" in their offspring; disproven by Mendel's work.

BLITZ EBI's ultra-fast protein database search which uses the MPsearch algorithm.

BLOCKS A database of ungapped multiple alignments for protein/peptide families in PROSITE.

blocks database A public database of protein patterns that correspond to the most highly conserved regions in proteins.

BLOSUM Blocks Substitution Matrix. A substitution matrix in which scores for each position are derived from observations of the frequencies of substitutions in blocks of local alignments in related proteins. Each matrix is tailored to a particular evolutionary distance. In the BLOSUM62 matrix, for example, the alignment from which scores were derived was created using sequences sharing not more than 62% identity. Sequences more identical than 62% are represented by a single sequence in the alignment so as to avoid over-weighting, closely related family members.

blotting (Blots) The process of transferring DNA, RNA, or proteins to a solid support (usually a sheet of nitrocellulose paper) for hybridization after it has been separated by electrophoresis. Blots are named according to the material that is analysed. A Southern blot examines DNA which has been cut with restriction enzymes and probed with radioactive DNA. The northern blot analyses RNA which is probed with radioactive DNA or RNA. Western blots examine proteins which are probed with radioactive or enzymatically-tagged antibodies.

blunt end An end of a double-stranded DNA molecule where both strands terminate at the same nucleotide position with no single-stranded extension.

blunt end ligation The joining of DNA fragments that contain no overhang at either end and consequently no DNA bases available for hybridization.

boid An artificial life program, developed by Craig Reynolds in 1986, simulates the flocking behaviour of birds. It is an autonomous agent that behaves like a simplified bird, but will display flocking patterns in the presence of other boids.

Bone Marrow Transplantation (BMT) A medical procedure to replenish the soft tissue within bones that produces new blood cells.

Boolean A variable that has a value of either 1 ("true" or "yes") or 0 ("false" or "no").

Boolean network A set of (*N*) Boolean variables, in which each variable is a function of (*K*) other variables in the set.

bootstrap analysis A statistical method that is often used to estimate the reproducibility of specific features of phylogenetic trees.

bottleneck A severe reduction in the size of a population, which limits the pool of alleles that are passed on to future generations of the species.

bottom-up A description that uses lower level details to explain higher level patterns.

box A short DNA sequence adjacent to or residing within a gene that performs a regulatory function, e.g. TATA box.

Boyce–Codd Normal Form (BCNF) A database relation in which every determinant is a candidate key.

branch The graphical representation of an evolutionary relationship in a phylogenetic tree.

branch migration A step in the Holliday model for homologous recombination, involving exchange of polynucleotides between a pair of recombining double-stranded DNA molecules.

BRCA1/BRCA2 The first breast cancer genes to be identified. Mutated forms of these genes are believed to be responsible for about half the cases of inherited breast cancer, especially those that occur in younger women. Both are tumour-suppressor genes.

brown noise The sonic equivalent to Brownian motion; a form of randomness that is the result of cumulatively adding white noise, to yield a random walk pattern. This is useful in formation of self-evolving subsentient and sentient soundscapes.

browser A program used to access sites on the World Wide Web.

BSML (*See* Bioinformatic Sequence Markup Language).

bucket brigade algorithm A learning algorithm that is a method for adjusting the strengths of the classifiers of a classifier system.

bundle An array of alpha-helices each oriented roughly along the same (bundle) axis.

byte An amount of computer memory that holds a single text character, usually eight bits.

C

C++ A general-purpose programming language. It is regarded as a middle-level language, as it comprises a combination of both high-level and low-level language features. It was developed by Bjarne Stroustrup in 1979 at Bell Labs as an enhancement to the C programming language and originally named as "C with Classes" and then it was renamed to C++ in 1983.

Caenorhabditis elegans A small worm that is used as a model organism to study genetics and development.

cancer Any of a large number of diseases characterized by the uncontrolled proliferation of cells.

candidate gene A gene, located in a chromosome region suspected of being involved in a disease, whose protein product suggests that it could be the disease gene in question.

candidate key A combination of attributes that can be used to uniquely identify a database record without any extraneous data.

canonical complexity class A complexity class defined by logarithmic, polynomial, and exponential bounds on time and space, for deterministic and non-deterministic machines.

Cantor set A fractal set composed of uncountably many dust-like points, that has zero measure (meaning that the sum width of all points is 0).

cap A complex structure at the 5′ termini of most eukaryotic mRNA molecules, having a 5′–5′ linkage instead of the usual 3′–5′ linkage.

cap-binding complex The complex, also called eIF-4F and comprising the initiation factors eIF-4A, eIF-4E and eIF-4G, which makes the initial attachment to the cap structure at the beginning of the scanning phase of eukaryotic translation.

capillary array Gel-filled silica capillaries used to sequence DNA.

capping Attachment of a cap to the 5′ end of pre-mRNA in eukaryotes whereby a GTP is added to the molecule via a 5′–5′ triphosphate bond.

capsid The protein coat that surrounds the DNA or RNA genome of a virus.

cap site A DNA binding site for the catabolite activator protein; in DNA, the site where transcription starts; in RNA, the site that is capped during the process of mRNA maturation.

carboxyl group The –COOH functional group, acidic in nature, found in all amino acids.

carboxyl terminus The COOH end of a polypeptide.

carcinogen A physical or chemical agent that causes cancer by making changes in a cell's DNA.

carcinoma A type of cancer in which tumours form in the epithelial tissue, the tissue forming the outer layer of the body surface and lining the digestive tract and other hollow structures.

cardinality The number of rows (or tuples) contained in a database table.

carrier An individual heterozygous for a recessive trait.

carrying capacity The maximum number of individuals of a given species that can be sustained in a defined habitat.

Case-Based Reasoning (CBR) An approach to computer reasoning that uses knowledge from a library of similar cases, rather than by accessing a knowledge base containing more generalized knowledge, such as a set of rules.

CAS Registry Number (CAS-RN) A unique accession number assigned by the Chemical Abstracts Service, a division of the American Chemical Society.

Catabolite Activator Protein (CAP) A regulatory protein that binds to various sites in a bacterial genome and activates transcription initiation at downstream promoters.

catabolite repression A mechanism by which metabolic products (catabolites) can determine the degree of expression of enzymes in their metabolic pathway.

catalyst A compound that lowers the energy necessary to activate a chemical reaction, without being consumed or altered by itself. Proteins are the primary catalysts in the cell, but RNA can act as a catalyst as well.

catastrophism The idea that Earth has been affected in the past by sudden, short-lived violent events, possibly world-wide in scope.

causal reasoning A form of reasoning based on following from cause to effect.

cDNA capture Repeated hybridization probing of a pool of cDNA to obtain a subpool enriched in certain sequences.

cDNA (complementary DNA) A DNA strand copied from an mRNA using reverse transcriptase. A cDNA library represents all of the expressed DNA in a cell. The single-stranded form of cDNA is frequently used as a probe in the preparation of a physical map of a genome. cDNA is preferred for sequence analysis because the introns found in DNA are removed in post- transcription DNA $\rightarrow$ mRNA $\rightarrow$ cDNA.

cDNA library A collection of DNA sequences generated from mRNA sequences. This type of library contains only protein-coding DNA (genes) and does not include any non-coding DNA. It is a set of DNA fragments prepared from the total mRNA obtained from a selected cell, tissue or organism.

cell The basic structural and functional unit of life. It is the simplest unit that can exist as an independent living system. Cells are classified into two: either prokaryotic and eukaryotic.

cell cycle The series of events that occur in a cell between one division and the next. It is the growth cycle of a cell; in eukaryotes, it is subdivided into G_1 (gap 1), S (DNA synthesis), G_2 (gap 2), and M (mitosis).

cell cycle checkpoint A period before entry into S phase or M phase of the cell cycle, a key point at which regulation is exerted.

cell-free protein-synthesizing system A cellular extract that contains all the components needed for protein synthesis and performs translation of added mRNA molecules.

Cellular Automaton (CA) A discrete dynamical system that is composed of an array of cells, each of which behaves like a finite-state automaton.

cellular bioinformatics The lesser developed branch of bioinformatics that focuses on the understanding of the functioning living cell. As such it has to integrate DNA, mRNA, protein and metabolic data. Because of the complexity of the problem, it also needs to invoke mathematical modelling. The branch of cellular bioinformatics that focuses on understanding on the basis of all the known experimental data is also called as computational biochemistry.

Centre for Bioinformatics and Computational Biology (CBCB) A program of the National Institute of General Medical Sciences (NIGMS).

centiMorgan (cM) The unit of measurement for distance and recombination frequency on a genetic map. Formally, the length (number of bases) that have a 1% probability of participating in mixing of genes. For humans, the average length of a cM is one million base pairs (or 1 megabase, Mb).

central dogma The path of information flow in DNA organisms (DNA → RNA → proteins).

Central Limit Theorem (CLT) A fundamental theorem of probability and statistics that states the conditions under which the distribution of a sum of independent random variables is approximated by the normal distribution.

centriole Paired cellular organelle which functions in the organization of the mitotic spindle during cell division in eukaryotes.

centroid method A clustering approach in which one first identifies a set of "centroids" or "centres of mass" for the given data points.

centromere A specialized chromosome region to which spindle fibres attach during cell division.

chain termination method A DNA sequencing method that involves enzymatic synthesis of polynucleotide chains that terminate at specific nucleotide positions.

chain termination mutation A mutation that leads to the premature termination of the polypeptide chain specified by the mutant gene.

chaos Irregular motion of a dynamical system that is deterministic, sensitive to initial conditions, and impossible to predict in the long term with anything less than an infinite and perfect representation of analog values.

chaperonin A multi-subunit protein that forms a structure that aids the folding of other proteins.

character In phylogenetics, characters are homologous

features in different organisms. The exact condition of that feature in a particular individual is the character state.

character state One of at least two alternative forms of a character used in phylogenetic analysis.

CHARMM (Chemistry at HARvard Macromolecular Mechanics) A type of molecular mechanics force field used to predict the 3D structure of proteins.

chemical degradation sequencing A DNA sequencing method that involves the use of chemicals that cut DNA molecules at specific nucleotide positions.

chiasma The point (literally, a cross) at which a genetic recombination event has occurred, visible as a crossover in meiotic prophase.

chi form An intermediate structure seen during recombination between DNA molecules.

chimera An organism composed of two or more genetically different cell types.

chimeric clone A cloning artifact created by a foreign gene being inserted into a vector in an incorrect orientation resulting in the expression of a protein consisting of a fusion of two different gene products.

chimeric protein A protein encoded by a gene that contains regions from different genes. Also, an artificial protein, derived through genetic engineering.

chi site A repeated nucleotide sequence in the *Escherichia coli* genome that is involved in the initiation of homologous recombination.

chloroplast One of the photosynthetic organelles of an eukaryotic cell.

chloroplast genome The circular genome present in the chloroplasts of a photosynthetic eukaryotic cell.

Chomsky hierarchy A containment hierarchy of four classes of formal grammars that generate languages of increasing complexity.

Chorionic Villus Sampling (CVS) A technique used in antenatal diagnosis where a small amount of placental tissue is removed at about 8–9 weeks of gestation by transcervical aspiration.

chromat Data file output from most popular DNA sequencers. Chromat files consist of the fluorescent traces generated by the sequencer for each of the four chemical bases A, C, G, and T, together with the sequence and measures of the error in the traces at each sequence position.

chromatid Each of the two copies produced by chromosome replication, or each of the two DNA strands that comprise the chromosome.

chromatin The complex of DNA and histone proteins found in chromosomes.

chromatin remodelling A process that results in repositioning of nucleosomes on a eukaryotic DNA molecule.

Chromatin Remodelling Machine (CRM) The set of enzymes that carry out chromatin remodelling.

chromosome A self-replicating genetic structure of cells, formed from DNA and protein, found in the cell nucleus. It is a collection of DNA and protein which organizes the human genome. Each human cell contains 23 sets of chromosomes; 22 pairs of autosomes (non-sex-determining chromosomes) and one pair of sex-determining chromosomes. The human genome within the 23 sets of chromosomes is made of approximately 30,000 genes which are built from over 3 billion base pairs.

While eukaryotic chromosomes are complex sets of proteins and DNA, prokaryotic chromosomal DNA is circular with the entire genome on a single chromosome.

chromosome jumping A cloning method which can be used to isolate clones at a desired distance (in the range of 100–300 kb) along a chromosome from the starting clone.

chromosome map A diagram showing the locations and relative spacing of genes along a chromosome.

chromosome painting Attachment of certain fluorescent dyes to the targeted parts of the chromosome.

chromosome regions p and q During cell division, the molecules that compose chromosomes suffer a condensation process, and forms a small compact complex. In diploid organisms, homologous chromosomes get attached to each other by the centromere. The centromere divides each chromosome into two regions: the smaller one is called p region and the bigger one is called q region.

chromosome theory of inheritance Theory that holds that chromosomes are the cellular components that physically contain genes; proposed in 1903 by Walter Sutton and Theodore Boveri.

chromosome walking A technique that can be used to construct a clone contig by identifying overlapping fragments of cloned DNA.

circuit A path that begins and ends at the same vertex.

cis The arrangement of two sequences or genes on the same chromosome.

clade A group of monophyletic organisms or DNA sequences that include all of those in the analysis that are descended from a particular common ancestor.

cladistics A phylogenetic approach that stresses the importance of understanding the evolutionary relevance of the characters that are studied. A particular systematic approach to building evolutionary trees and interpreting their meaning.

cladogenesis The process of evolutionary branching of basic forms into specialized forms by differentiation, specialization and adaptation.

cladogram A graphical representation that portrays or attempts to portray the evolutionary relationships among a number of populations, species, or higher taxa.

classifier A rule that is part of a classifier system and has a condition that must be matched before its message (or action) can be posted (or effected). The strength of a classifier determines the likelihood that it can outbid other classifiers if more than one condition is matched.

classifier system An adaptive system similar to a post-production system that contains many "if then" rules called classifiers. The state of the environment is encoded as a message by a detector and placed on the message list from which the condition portion of the classifiers can be matched. "Winning" classifiers can then post their own messages to the message list, ultimately forming a type of computation that may result in a message being translated into an action by an effector. The strengths of the classifiers are modified by the bucket brigade algorithm, and new rules can be introduced via a genetic algorithm.

Cleavage and Polyadenylation Specificity Factor (CPSF) A protein that plays an ancillary role during polyadenylation of eukaryotic mRNAs. It is also called as cleavage stimulation factor (CStF).

client A computer connected to a network that does not store all the data or software it uses, but retrieves it across the network from another computer that acts as a server.

client-server architecture A computer network architecture that places commonly used resources on centrally accessible computer servers, which can be retrieved as they are needed across the network by client computers on the network.

clone An exact copy of biological material such as a DNA segment (a gene or other region), a whole cell, or a complete organism. It is a population of genetically identical cells or DNA molecules.

clone contig A set of cloned-DNA fragments overlapping in such a way to provide unbroken coverage of a contiguous region of the genome.

clone-contig approach A genome sequencing strategy in which the molecules to be sequenced are broken into manageable segments, each up to a few Mb in length, which are sequenced individually.

clone fingerprinting One of several techniques that compares cloned DNA fragments in order to identify ones that overlap.

cloning The technique used to produce copies of a piece of DNA. A DNA fragment that contains a gene of interest is inserted into the genome of a virus or plasmid which is then allowed to replicate.

cloning vector A piece of DNA from any foreign body which is grafted into a host DNA strand that can then self replicate. Vectors are used to introduce foreign DNA into host cells for the purpose of manufacturing large quantities of the new DNA or the protein that the DNA expresses.

closed circular DNA A DNA molecule that has no free 5′ end or 3′ end. Plasmids and certain DNA viruses have this structure.

closed-loop control A completely automated system control method in which no part of the control system need be given over to humans.

closed-promoter complex The structure formed during the initial step in assembly of the transcription initiation complex.

cloverleaf A two-dimensional representation of the structure of a tRNA molecule.

ClustalW A widely used general purpose algorithm for multiple sequence alignment of DNA or proteins. It is a general purpose program for multiple alignments of DNA and protein sequences developed by Thompson *et al.* in 1994.

cluster The grouping of similar objects in a multi-dimensional space. Clustering is used for constructing new features which are abstractions of the existing features of those objects. The quality of the clustering depends crucially on the distance metric in the space. In bioinformatics, clustering is performed on sequences, high-throughput expression and other experimental data. Clusters of partial or complete gene sequences can be used to identify the complete (contiguous) sequence and to better identify its function. Clustering expression data enables the researcher to discern patterns of co-regulation in groups of genes.

cluster analysis A process of assigning data points (e.g. sequences) into groups (clusters), usually starting from the pairwise distances between points.

Clusters of Orthologous Groups (COG) A phylogenetic classification of proteins encoded in complete genomes.

co-recursively enumerable (co-RE) The complement of a set that is recursively enumerable. For example, a set S of natural numbers is called recursively enumerable if there is a partial recursive function (synonymously, a partial computable function) whose domain is exactly S, meaning that the function is defined if and only if its input is a member of S.

co-repressor A small molecule that must be bound to a repressor protein before the latter can attach to its operator site.

coding regions All exon parts of a protein-coding gene that are ultimately translated. It is the portion of a genomic sequence bounded by start and stop codons that identifies the sequence of the protein being coded for by a particular gene.

coding RNA An RNA molecule that codes for a protein, that is, an mRNA.

co-dominance The relationship between a pair of alleles in which both contribute to the phenotype of a heterozygote.

codon A sequence of three nucleotides in messenger RNA that codes for a single amino acid; a sequence of three adjacent nucleotides that designates a specific amino acid or start/stop site for transcription. It is the set of three nucleotides along a strand of mRNA that determines (or codes) the amino acid placement during protein synthesis. The number of possible arrangements of these three nucleotides (or triplet codes) available for protein synthesis (4 bases)^3 = 64. Thus, each amino acid can be coded by up to 6 different triplet codes. Three triplet codes (UAA, UAG, UGA) specify the end of the protein. In the example below, three codons are shown.

— U C A — C G U — C A U —
— Ser — Arg — His —

codon–anticodon recognition The interaction between a codon on an mRNA molecule and the corresponding anticodon on a tRNA molecule.

codon bias The phenomenon where codons for a specific amino acid do not appear with equal frequency in the genes of a particular organism.

codon family The codons that specify the same amino acid and differ from each other only at the third position.

codon usage The observed usage of the codons in a particular gene, tissue, or organism.

coefficient of inbreeding The proportion of the genes of an individual in which both copies are identical by descent from a common ancestor.

coenzyme An organic non-protein molecule that does not bind the enzyme, but is required for the enzyme's function by acting as an intermedi-ate carrier of electrons, atoms, or groups of atoms.

co-evolution Evolution of two or more entities in response to one another.

cofactor An inorganic molecule required by an enzyme in order to function.

coincidental substitution The occurrence of two substitutions at the same nucleotide site in two homologous sequences.

cointegrate An intermediate in the pathway resulting in replicative transposition.

collagen The most abundant protein in mammals which is an extracellular, rod-shaped, protein containing three helical polypeptide chains, each nearly 1000 residues in length.

collagenase An enzyme that degrades collagen.

colinearity The exact correspondence between the DNA sequence of intron-less genes and the amino acid sequence of the encoded protein.

colony The cluster of bacteria that grow from a single bacterium on a nutrient agar dish after a period of incubation.

colony screen A technique to identify cells that contain a desired probe sequence.

column Database tables are composed of individual columns corresponding to the attributes of the object.

combinatorial chemistry The technology of synthesizing novel compounds using various molecular building blocks. It uses the chemical methods to generate all possible combinations of chemicals starting with a subset of compounds. The building blocks may be peptides, nucleic acids or small molecules. The libraries of compounds formed by this methodology are used to probe for new pharmaceutical reagents.

combinatorial library A collection or database of compounds used to assay a drug target.

combinatorial optimization A class of problems in which the number of candidate solutions is combinatorial in size.

command line A means of interacting with computer software by typing specific commands.

commensalism A symbiotic relationship in which one species benefits and the other is not affected.

Common Object Request Broker Architecture (CORBA) An open, vendor-independent architecture and infrastructure that allows computer applications to work together over networks. It is a technology specification (sometimes referred to as a wrapper) that uses an interface definition language (IDL, code which defines the properties of data modules or objects) and software (the Object Request Broker or ORB) to define how objects (self-contained modules of data or instructions) can share the characteristics needed to form a unified application.

community All species or populations living in the same area.

comparative bioinformatics The genome sequences from several chordates are being completed; the bioinformatics largely exists in the research community to discover the protein-coding potential of those genomes. However, the bioinformatics to elucidate gene regulation encoded in genomes and gene regulatory networks is not so developed. New bioinformatics, new model organism resources, new experimental approaches, and new collaborations are needed if the community is to understand the gene networks that help create phenotypes of interest. A research team at ORNL and the University of Tennessee are developing some needed bioinformatics.

The overall projects include 1) supplying several web services and collaborative bioinformatics that supports large consortia of experimental researchers and 2) developing comparative bioinformatics and new data mining environments that can ultimately help understand the nature and evolution of gene regulatory networks.

comparative genomics The study of a genome by comparing complete genomes, often by computational methods, to understand general principles of genome structure and function.

Comparative Molecular Field Analysis (CoMFA) A technique used to explore structure–activity relationships between proteins and ligands.

competent Referring to bacteria that have been treated, for example, by soaking in calcium chloride, so that their ability to take up DNA molecules is enhanced.

competition One of the biological interactions that can limit population growth; occurs when two species vie with each other for the same resource.

competitive exclusion Competition between species that is so intense that one species completely eliminates the second species from the area.

competitive release Process that occurs that when one of the two competing species is removed from an area, thereby releasing the remaining species from one of the factors that limit its population size.

compiler A program that translates a higher level language (C, FORTRAN, Pascal), whose commands that the computer cannot understand directly, into a machine-language file that the computer can run.

complement A set composed of all elements that are not members of another set.

complementarity The sequence-specific or shape-specific recognition that occurs when two or more molecules bind together. DNA forms double-stranded helices because the complementary orientation of the bases in each strand facilitate the formation of the hydrogen bonds which hold the strands together.

complementary Refers to the two nucleotides or nucleotide sequences that are able to form base pairs with one another.

Complementary Determining Region (CDR) The hypervariable regions of an antibody molecule, consisting of three loops from the heavy chain and three from the light chain, that together form the antigen-binding site.

complementary non-deterministic polynomial-time (co-NP) The set (or property) of problems with a yes/no answer where the complementary no/yes problem is in the set NP.

complementary sequence A sequence of bases that can form a double-stranded structure by matching base pairs.

complementation The ability of two different mutations to produce a wild-type phenotype in a double heterozygote.

complete A formal logical system in which all statements can be proved either true or false.

complete dominance The type of inheritance in which both heterozygotes and dominant homozygotes have the same phenotype.

complete graph A graph in which each vertex is connected to each of the others (with one edge between each pair of vertices).

complexity The level in difficulty in solving mathematically posed problems as measured by the time, number of steps or arithmetic operations, or memory space required (called time complexity, computational complexity, and space complexity, respectively).

In molecular genetics, used to describe a DNA molecule or a mixture of DNA molecules. It is the length of the sequence without including any sequence repetition.

complexity class A set of computational problems that share the same bounds on requirements for time and space.

complexity (of gene sequence) The term "low complexity sequence" may be thought of as synonymous with regions of locally biased amino acid composition. In these regions, the sequence composition deviates from the random model that underlies the calculation of the statistical significance (P-value) of an alignment. Such alignments among low complexity sequences are statistically but not biologically significant, i.e., one cannot infer homology (common ancestry) or functional similarity.

complex number A number that is the sum of a real number and an imaginary number.

Complex Pathway Simulator (COPASI) A software package for advanced modelling of biochemical systems; the successor of Gepasi.

complex system A collection of many simple, non-linear units that operate in parallel and interact locally with each other so as to produce emergent behaviour.

complex trait A trait that has a genetic component that does not follow strict Mendelian inheritance.

complex transposon A transposon flanked by two complete, independently transposable insertion sequences.

composite transposon A DNA transposon composed of a pair of insertion sequences.

compositional assimilation The accumulation of point mutations in a pseudogene that eventually obliterates its sequence similarity to the functional gene from which it has been derived and makes its nucleotide composition similar to neighbouring DNA sequences.

compound heterozygote The presence of two different mutant alleles at a particular gene locus, one on each chromosome of a pair, i.e., the human genome contains two copies of each gene, a paternal and a maternal allele. A mutation affecting only one allele is called heterozygous. A homozygous mutation is the presence of the identical mutation on both alleles of a specifc gene. However, when both alleles of a gene harbour mutations, but the mutations are different, these mutations are compound heterozygous, also called as genetic compound. That is, for a locus A with normal allele A1 and mutant alleles A2 and A3, a compound heterozygote will be A2/A3.

compressible Having a description that is smaller than the object itself.

computable Expressible as yes/no question that a computer can answer within a finite period of time.

computational biology A field incorporating computer science and biology. It is the development and application of data-analytical and theoretical methods, mathematical modelling and computational simulation techniques to the study of biological, behavioural, and social systems.

concatamer A DNA molecule made up of linear genomes linked head-to-tail.

conceptual biology The iterative process of analysing existing facts and models available in published literature to generate new hypotheses in which predictions are formulated in testable terms, and then search for relevant information among published reports of experiments that may have had a different purpose altogether.

concerted evolution Maintenance of homogeneity of nucleotide sequences among members of a gene family in a species, although the nucleotide sequences change over time.

CONCORD A software program that calculates the 3D structure of small molecules.

conditional fixation time The time which takes a mutant allele to achieve fixation in a population.

conditional-lethal mutation A mutation that results in a cell or organism able to survive only under permissive conditions.

conditional probability The probability, written Pr(A/B), of event A given that event B has occurred.

Confidence Interval (CI) A range of values within which a result is expected to fall with a specific probability.

confidentiality In genetics, the expectation that genetic material and the information gained from testing that material will not be available without the donor's consent.

configuration The ordering and description of all parts of a software or database system.

conformation The precise three-dimensional arrangement of atoms and bonds in a molecule which describe its geometry and, hence, its molecular function.

congenic Nearly identical strains of an organism; they vary only at a single locus.

congenital Present at birth, but not necessarily inherited.

conjugation Transfer of DNA between two bacteria that come into physical contact with each other.

conjugation mapping A technique for mapping bacterial genes by determining the time it takes for each gene to be transferred during conjugation.

conjunction The Boolean "and" function.

connected graph A graph is connected if there is a path connecting every pair of vertices.

connectionism The study of the theory and application of artificial neural networks. More generally, it's an approach that emphasizes the connections among concepts, rather than their symbolic meaning.

connectivity The amount of interaction in a system, the structure of the weights in an artificial neural network, or the relative number of edges in a graph.

consanguineous mating A mating in which male and female are related by descent.

consensus map The location of all consensus sequences in a series of multiple aligned proteins or polynucleotides.

consensus sequence A derived base sequence that represents a family of similar sequences. It is a single sequence delineated from an alignment of multiple constituent sequences that represents a "best fit" for all those sequences. A "voting" or other selection procedure is used to determine which residue (nucleotide or amino acid) is placed at a given position in the event that not all of the constituent sequences have the identical residue at that position. Thus it is the most commonly occurring amino acid or nucleotide at each position of an aligned series of proteins or polynucleotides.

conservation Changes at a specific position of an amino acid or, less commonly, DNA sequence that preserve the physico-chemical properties of the original residue.

conservative replication A hypothetical mode of DNA replication in which one daughter double helix is made up of the two parental polynucleotides and the other is made up of two newly synthesized polynucleotides.

conservative substitution The substitution of an amino acid by another with similar chemical properties.

conservative system A dynamical system that preserves the volume of its state space under motion and, therefore, does not display the types of behaviour found in dissipative systems.

conservative transposition The movement (transposition) of a transposable element from one position in the genome to another without replication of the element.

conserved sequence A sequence within DNA or protein that is consistent across species or has remained unchanged within the species over its evolutionary period.

consistency 1. The requirement that only valid data will be written to a database. 2. A property of formal logical systems in which all statements are either true or false.

consomic strain A strain of organism in which a single, full-length chromosome from one inbred strain, the donor strain, has been transferred onto the genetic background of a second strain, termed "the host strain", by repeated backcrossing.

constitutive heterochromatin Chromatin that is permanently in a compact form.

constitutive mutation A mutation that results in continuous expression of a gene or set of genes that normally would be subject to regulatory control.

constitutive synthesis (expression) Synthesis of mRNA and protein at an unchanging or constant rate regardless of a cell's requirements.

Context-Free Grammar (CFG) A grammar in which every production rule is of the form: $V \to w$ where, V is a non-terminal symbol and w is a string consisting of terminals and/or non-terminals.

contig A group of cloned (copied) pieces of DNA representing overlapping regions of a particular chromosome. A DNA sequence that overlaps with another contig. The length of contiguous sequence assembled from partial, overlapping sequences, generated from a "shotgun" sequencing project. Contigs are typically created computationally, by comparing the overlapping ends of several sequencing reads generated by restriction enzyme digestion of a segment of genomic DNA. The creation of contigs in the presence of sequencing errors, ambiguities and the presence of repeats is one of the most computationally challenging aspects of the role of bioinformatics in genome analysis.

contig map(s) A map depicting the relative order of a linked library of small overlapping clones representing a complete chromosomal segment. It is the representation of the structure of contiguous regions of the genome (contigs) by specifying overlap relationships among a set of clones.

contiguous gene syndrome A syndrome caused by a deletion which has removed one copy of a number of genes which are close together in the genome.

contingent strategy A strategy that depends on behaviour in previous plays of the (repeated) game.

continuous Having a real-number value, analog. Not discrete.

continuous variation Variation that occurs when the phenotypes of traits controlled by a single gene cannot be sorted into two distinct phenotypic classes, but rather fall into a series of overlapping classes.

Contour Clamped Homogeneous Electric Fields (CHEF) An electrophoresis method used to separate large DNA molecules.

control Exerting actions to manipulate a system or environment in a goal-seeking manner.

controlled vocabulary A vocabulary that contains specific words that are applied consistently to all entries in a database.

conventional pseudogene A gene that has become inactive because of the accumulation of mutations.

convergence The property of approaching a stable solution. It is the endpoint of any algorithm that uses iteration or recursion to guide a series of data-processing steps. An algorithm is usually said to have reached convergence when the difference between the computed and observed steps falls below a pre-defined threshold.

convergent evolution The independent evolution of similar genotypes or phenotypes to yield similar character states.

convergent substitution The substitution of two different nucleotides by the same nucleotide at the same nucleotide site in two homologous sequences.

cooperative game A game in which joint-action agreements, pre-game contracts, etc. are enforceable.

CORBA *See* Common Object Request Broker Architecture.

core enzyme The version of *E. coli* RNA polymerase, subunit composition a2_W, that carries out RNA synthesis but is unable to locate promoters efficiently.

core octamer The central component of a nucleosome, made up of two subunits each of histones H2A, H2B, H3 and H4, around which DNA is wound.

core promoter The position within a eukaryotic promoter where the initiation complex is assembled.

cosmid An artificially constructed cloning vector containing the *cos* gene of the phage lambda. It allows the insertion of long fragments of DNA (up to 50 kb) that can be replicated inside *E. coli* bacteria.

co-transduction Transfer of two or more genes from one bacterium to another via a transducing phage.

co-transformation Uptake of two or more genes on a single DNA molecule during transformation of a bacterium.

countably infinite Having the same number of objects as the set of natural numbers.

covalent bond A chemical bond created by the sharing of electrons between atoms.

CpG island A DNA segment consisting of dimers of cytosine (C) and guanine (G).

CPK model A style of displaying 3D protein structures in a space-filling model, where oxygen is coloured red, nitrogen blue, carbon black, and hydrogen is white.

cREB cAMP Response Element Binding. A protein which is a transcription factor that binds to certain DNA sequences called cAMP response elements (CRE) and thereby increases or decreases the transcription, and thus the expression, of certain genes.

crossing over The exchange of DNA between chromosomes during meiosis. It is the interchange of two pieces of homologous chromosomes.

crossover The reciprocal exchange of genetic material between homologous chromosomes in meiosis. In genetic algorithms, an operator that splices information from two or more parents to form a composite offspring that has genetic material from all parents.

cryptic splice site A site, whose sequence resembles an authentic splice site, that might be selected instead of the authentic site during aberrant splicing.

cryptogene One of several genes in the trypanosome's mitochondrial genome which specify abbreviated RNAs that must undergo pan-editing in order to become functional.

crystal structure Term used to describe the high-resolution molecular structure derived by X-ray crytallographic analysis of protein or other biomolecular crystals.

csh The UNIX C shell, an environment where you can issue commands to the kernel.

CTD-Associated SR-like Protein (CASP) A protein that may play help to regulate splicing of GU–AG introns.

curated databases Often less complete than primary databases, but they have less redundancy and the added value of scientific annotation; therefore, a biologically significant sequence should be easier to find in such a database and of greater value. Naturally, the degree of redundancy and annotation in such a database depends on the experience, skills, aims, and devotion of its curators. The only proper way to curate databases is the way groups like those that developed OMIM [Online Mendelian Inheritance in Man], Swiss- Prot and most commercial databases have done it — that is, through making scientific judgments as data are cleaned up and merged.

cursor A database object used to traverse the results of an SQL query.

cut vertex A vertex that if removed (along with all edges incident to it) produces a graph with more connected components than the original graph.

C-value The characteristic amount of DNA in the haploid genome of a species.

C-value paradox The apparent lack of correlation between an organism's C-value and its level of morphological complexity.

cyanelle A photosynthetic organelle that resembles an ingested cyanobacterium.

cyanobacteria A phylum of bacteria that obtain their energy through photosynthesis, also known as blue-green algae-green bacteria or Cyanophyta. The name "cyanobacteria" comes from the colour of the bacteria. They are a significant component of the marine nitrogen cycle and an important primary producer in many areas of the ocean, but are also found in habitats other than the marine environment; particular cyanobacteria are known to occur in both freshwater and hypersaline inland lakes.

cybernetics The study of feedback control systems and their application.

cycle A path whose first vertex and last vertex are the same.

cyclic AMP A modified version of AMP in which an intramolecular phosphodiester bond links the 5´ and 3´ carbons.

cyclin A regulatory protein whose abundance varies during the cell cycle and which regulates biochemical events in a cell cycle-specific manner.

cyclobutyl dimer A dimer between two adjacent pyrimidine bases in a polynucleotide, formed by ultraviolet irradiation.

cys2His2 finger A type of zinc finger DNA-binding domain.

Cystic Fibrosis (CF) An autosomal recessive genetic disease involving a sticky build-up of mucus in the lungs (which makes breathing difficult and leads to infections), as well as pancreatic insufficiency (which leads to digestive problems).

cystine (Cys, C) A non-essential amino acid formed when two cysteine amino acid residues form disulphide bond.

cytogenetic location The region (or point) in the cytogenetic banding pattern within (or at) which a gene (or cloned DNA fragment) is thought to lie.

cytogenetic map The visual appearance of a chromosome when stained and examined under a microscope.

cytogenetics The study of the physical appearance of chromosomes.

cytokinesis The division of the cytoplasm during cell division.

cytoplasm The viscous semi-liquid inside the plasma membrane of a cell; contains various macromolecules and organelles in solution and suspension. It is the medium of the cell between the nucleus and the cell membrane.

cytoplasmic trait A genetic characteristic in which the genes are found outside the nucleus, in chloroplasts or mitochondria.

Cytosine (C) A pyrimidine base found in DNA and RNA that base-pairs with guanine.

cytoskeleton A three-dimensional network of microtubules and filaments that provides internal support for the cells, anchors internal cell structures, and functions in cell movement and division.

D

dark repair A type of nucleotide excision repair process that corrects cyclobutyl dimers.

D-arm A feature in the tertiary structure of transfer RNA. It is composed of the two D stems (four base pairs each; 10–13 and 22–25) and the D-loop. The D-loop contains the base dihydrouracil. The D-loop's main function is that of recognition. It is widely believed that it will act as a recognition site for aminoacyl-tRNA synthetase which is an enzyme involved in the aminoacylation of the tRNA synthetase which is an enzyme involved in the aminoacylation of the tRNA molecule. The D stem is also believed to have a recognition role although this is yet to be proved.

Darwinism A theory of evolution, proposed by Charles Darwin, that combines the variation of inheritable traits with natural selection.

data Information manipulated by a computer program.

database (db) A collection of information organized into interrelated tables of data and specifications of data objects. It is a file system by which data gets stored following a logical process of collection of data in machine-readable form, which can be manipulated by software to appear in varying arrangements and subsets.

Genetic information is stored in different ways in different databases, which makes it hard to compare their holdings. So while computational biologists are trying to improve the quality of the databases, they are also working to build bridges between them. So far, they have had only limited success. Each database has its own website with unique navigation tools and data storage formats that make searching difficult. Programs can't easily recognize data that are not stored in a uniform way.

data cleaning A process whereby automated or semi-automated algorithms are used to process experimental data, including noise, experimental errors and other artifacts, in order to generate and store high-quality data for use in subsequent analysis. Data cleaning is typically required in high-throughput sequencing where compression or other experimental artifacts limit the amount of sequence data generated from each sequencing run or "read".

data mining The use of automated data analysis techniques to uncover previously undetected relationships among data items. It is the ability to query very large databases in order to satisfy a hypothesis ("top-down" data mining) or to interrogate a database in order to generate new hypotheses based on rigorous statistical correlations ("bottom-up" data mining).

Data model standards group An effort that underway to standardize domain-specific analytical data models which help to integrate public data with proprietary data across all life science domains in an enterprise.

data processing The systematic performance of operations upon data such as handling, merging, sorting, and computing. The semantic content of the original data should not be changed, but the semantic content of the processed data may be changed.

data structure The hierarchy of data types in a program.

data type A description of a variable, defined by two properties: a domain, which is the set of values that belong to that type, and a set of operations, which defines the behaviour of that type.

data warehouse A database whose data may be accessed often, but is seldom changed. It is the vast arrays of heterogeneous (biological) data, stored within a single logical data repository, that are accessible to different querying and manipulation methods.

dbEST A division of the GenBank database that contains sequence data and other information on "single-pass" cDNA sequences, or expressed sequence tags (ESTs), from a number of organisms.

dbGSS The GSS division of GenBank is similar to the EST division, with the exception that most of the sequences are genomic in origin, rather than cDNA (mRNA). It should be noted that two classes (exon-trapped products and gene-trapped products) may be derived via a cDNA intermediate. Although dbGSS sequences are incorporated into the GSS Division of GenBank, annotation in dbGSS is more comprehensive and includes detailed information about the contributors, experimental conditions, and genetic map locations.

dbSTS The division of NCBI's GenBank resource that contains sequence and mapping data on short genomic landmark sequences or sequence tagged sites (STS).

deadenylation-dependent pathway A process for degradation of eukaryotic mRNAs that is initiated by removal of the poly(A) tail.

deadenylation-independent pathway A process for degradation of eukaryotic mRNAs that is initiated by the presence of an internal termination codon.

deaminating agent A mutagen that acts by removing amino groups from nucleotide bases.

decision problem A problem in which all questions are of the form: "Is the object a member of the set?" and all answers are either "yes" or "no."

Decision Support System (DSS) A computer application that enhances or assists an individual's or organization's ability to make decisions.

decision tree A method of representing knowledge that structures decisions in a hierarchical, tree-like fashion.

deconvolution Mathematical procedure to separate out the overlapping effects of molecules such as mixtures of compounds in a high-throughput screen, or mixtures of cDNAs in a high density array.

deduction A method of logical inference. Given a cause, deduction infers all logical effects that might arise as a consequence.

deductive database A database that contains both facts (often in the form of a relational database) and rules for reasoning (often in logic programming) so that new facts can be dynamically generated from stored facts.

degenerate Redundant referring to the fact that the genetic code has more than one codon for most amino acids.

degenerate code A genetic code in which the number of sense codons is larger than the total number of amino acids and some amino acids are specified by more than one codon.

degradosome A multienzyme complex that degrades (breaks down) bacterial mRNAs.

degree The number of edges that end at a specific vertex.

delayed-onset mutation A mutation whose effect is not apparent until a relatively late stage in the life of the mutant organism.

deleterious mutation A mutation that lowers the fitness of its carriers.

deletion The loss of a chromosome segment without altering the number of chromosomes.

deletion mapping A description of a specific chromosome that uses defined mutations—specific deleted areas in the genome—as "biochemical signposts," or markers for specific areas. In this process, different deletions in a region of DNA are created and used to map the functionally critical areas of that DNA, e.g. the minimal region of DNA required for a test promoter can be ascertained by systematic deletions in the region of interest.

deletion mutation A mutation resulting from loss of one or more nucleotides from a DNA sequence.

delta rule The perceptron learning rule that specifies that the weight changes should be proportional to the product of a weight's input and the error (or delta) term for the perceptron.

denaturation Breakdown by chemical or physical means of the noncovalent interactions, such as hydrogen bonds, that maintain the secondary and higher levels of structure of proteins and nucleic acids.

dendrogram A graphical procedure for representing the output of a hierarchical clustering method. A dendrogram is strictly defined as a binary tree with a distinguished root, that has all the data items at its leaves. Conventionally, all the leaves are shown at the same level of the drawing. The ordering of the leaves is arbitrary, as is their horizontal position. The heights of the internal nodes may be arbitrary, or may be related to the metric information used to form the clustering.

de novo methylation Addition of methyl groups to new positions on a DNA molecule.

deoxyribonuclease An enzyme that breaks phosphodiester bonds in a DNA molecule.

deoxyribonucleic acid (DNA)
See DNA.

deoxyribonucleotide A monomer or single unit of DNA or deoxyribonucleic acid. Each deoxyribonucleotide is comprised of three parts: A nitrogenous base, a deoxyribose sugar, and one or more phosphate groups. The nitrogenous base is always bonded to the $1'$ carbon of the deoxyribose, which is distinguished from ribose by the presence of a proton on the $2'$ carbon rather than on – OH group. The phosphate groups bind to the $5'$ carbon of the sugar. When deoxyribonucleotides polymerize to form DNA, the phosphate group from one nucleotide will bond to the $3'$ carbon on another nucleotide, forming a phosphodiester bond via dehydration synthesis. New nucleotides are always added to the $3'$ carbon of the last nucleotide, so synthesis always proceeds from $5'$ to $3'$.

deoxyribose A five-carbon sugar lacking a hydroxyl group on position 2 (β-D-2-deoxyribose) which is used in the construction of DNA.

Depth-First Search (DFS) A common recursive algorithm used to explore a graph.

depth of library The average number of times any sequence, originally present in a single copy in the genome, will be represented in a genomic library.

derivative An expression that characterizes how much a function's output changes as the input is varied.

derived character state A character state that evolved in a recent ancestor of a subset of organisms in the group being studied.

descriptor Information about a sequence or a set of sequences, whose scope depends on its placement in a record.

determinant A quantity of a matrix that characterizes the amount of expansion or contraction that the matrix inflicts on a vector when that vector is multiplied by the matrix.

deterministic Permitting at most one next move at any step in a computation; changing in a non-random manner, such that the next state of a system depends only on the prior states of the system and/or its environment.

deterministic algorithm An algorithm whose behaviour can be completely predicted from the input.

Deterministic Finite Automaton (DFA) A finite-state machine that recognizes regular languages, it has exactly one transition for each given symbol and state.

deterministic process A process whose outcome of which can be predicted exactly from knowledge of initial conditions.

development A coordinated series of transient and permanent changes that occurs during the life history of a cell or organism.

diagonal matrix A matrix whose non-diagonal cells have values of 0.

diauxie The phenomenon whereby a bacterium, when provided with a mixture of sugars, uses up one sugar before beginning to metabolize the second sugar.

dicentric A structurally abnormal chromosome with two centromeres.

dideoxynucleotide A modified nucleotide that lacks the 3′-hydroxyl group and so terminates strand synthesis when incorporated into a polynucleotide.

diff A UNIX program for comparing two files.

difference equation An equation that describes how something changes in discrete time steps.

differential centrifugation A common procedure in microbiology and cytology used to separate certain organelles from whole cells for further analysis of specific parts of cells. In the process, a tissue sample is first homogenized to break the cell membranes and mix up the cell contents. The homogenate is then subjected to repeated centrifugations, each time removing the pellet and increasing the centrifugal force. Finally, purification may be done through equilibrium sedimentation, and the desired layer is extracted for further analysis.

differential equation A description of how something continuously changes over time.

differentiation The developmental process by which an unspecialized cell undergoes a progressive change to a more specialized cell. In mathematics, it is the act of calculating a derivative; the inverse operation of calculating an integral.

diffusion-limited aggregation (DLA) A type of stochastic fractal formed by particles floating in a random manner until they stick to something solid.

digestion The cutting of a double-stranded DNA by a restriction endonuclease.

dihybrid cross A sexual cross in which one follows the inheritance of two pairs of alleles.

dimer A protein or other structure composed of two subunits. It is a composite molecule formed by the binding of two molecules.

diploid A nucleus (or cell) that has two copies of each chromosome.

Directed Acyclic Graph (DAG) *See* acyclic graph.

directed evolution A set of experimental techniques used to obtain novel genes with desired characteristics.

directed graph A graph whose edges have direction, typically represented as arrows.

directed mutagenesis Alteration of DNA at a specific site and its re-insertion into an organism to study any effects of the change.

directed sequencing Successively sequencing DNA from adjacent stretches of chromosome.

directed shotgun approach A genome sequencing strategy that combines random shotgun sequencing with a genome map, the latter used to aid assembly of the master sequence.

directional selection A selective process that changes the frequency of an allele in a specific direction, either toward fixation or towards elimination.

direct readout The recognition of a DNA sequence by a DNA-binding protein that makes contacts with the outside of a double helix.

direct repair A DNA repair system that acts directly on a damaged nucleotide.

direct repeat A nucleotide sequence that is repeated twice or more in a DNA molecule.

discontinuous gene A gene that is split into exons and introns.

discontinuous variation A variation occurs when the phenotypes of traits controlled by a single gene can be sorted into two distinct phenotypic classes.

discrete Taking only non-continuous values, such as Boolean values or integers.

disease-associated genes Alleles carrying particular DNA sequences associated with the presence of disease.

disjunction In bioinformatics, it is the Boolean or function. In molecular biology, it is the separation of homologous chromosomes during meiosis or the separation of complementary chromatids during mitosis.

dispersive replication A hypothetical mode of DNA replication in which both polynucleotides of each daughter double helix are made up partly of parental DNA and partly of newly synthesized DNA.

displacement replication A mode of replication which involves continuous copying of one strand of the helix, the second strand being displaced and subsequently copied after synthesis of the first daughter strand has been completed.

disruptive selection A process of natural selection that favours individuals at both extremes of a phenotypic range.

dissipative dynamical system A dynamical system whose internal friction deforms the structure of its attractor, which allows motion such as fixed points, limit cycles, quasiperiodicity and chaos.

distance A real-valued function applied to two points in a metric space.

distance matrix A table showing the evolutionary distances between all pairs of nucleotide sequences in a dataset. This method used to present the results of the calculation of an optimal pairwise alignment score. The matrix field (i, j) is the score assigned to the optimal alignment between two residues (up to a total of i by j residues) from the input sequences. Each entry is calculated from the top-left neighbouring entries by way of a recursive equation.

distance measure A function that associates a non-negative numeric value with a pair of biosequences. The shorter the distance between the sequences (i.e., the lower the number), the greater the similarity. For example, the distance between leu and ala is small while the distance between leu and arg is large.

distance method A rigorous mathematical approach to alignment of nucleotide sequences.

distributed annotation system A client-server system in which a single client integrates information from multiple servers. It allows a single machine to gather up genome annotation information from multiple distant websites, collate the information, and display it to the user in a single view. Little coordination is needed among the various information providers.

distributed computing An approach in which a computer system's data and programs are distributed across different computers on a network.

distributed sequence annotation The pace of human genomic sequencing has outstripped the ability of sequencing centres to annotate and understand the sequence prior to submitting it to the archival databases. Multiple third-party groups have stepped into the breach and are currently annotating the human sequence with a combination of computational and experimental methods. Their analytic tools, data models, and visualization methods are diverse, and it is self-evident that this diversity enhances, rather than diminishes, the value of their work.

disulphide bond Covalent link formed between the sulphur atoms of two different cysteine residues in a protein. It is important in maintaining the folded structure of a protein, and also for linking different proteins in a complex.

disulphide bridge A covalent bond linking cysteine amino acids on different polypeptides or at different positions on the same polypeptide.

diverge For algorithms or computers, to run without halting; for iterative systems, to reach a state such that all future states explode in size.

divergent evolution The divergence of a single interbreeding population or species into two or more descendant species.

dizygotic twins Twins arising by the fertilization of two eggs by two sperms.

D-loop (displacement loop) A loop formed when a part of a DNA strand is disloaded from a duplex molecule because of the partner strand's pairing with another molecule, i.e., a DNA structure where the two strands of a double-stranded DNA molecule are separated for a stretch and held apart by a third strand of DNA. The third strand has a base sequence which is complementary to one of the main strands and pairs with it, thus displacing the other main strand in the region. Within that region the structure is thus a form of triple-stranded DNA.

DNA adenine methylase (Dam) An enzyme involved in methylation of DNA in *E. coli*.

DNA bank A service that stores DNA extracted from blood samples or other human tissue.

DNA bending A type of conformational change introduced into a DNA molecule by a binding protein.

DNA-binding motif The part of a DNA-binding protein that makes contact with the double helix.

DNA-binding protein The proteins that are composed of DNA-binding domains and thus have a specific or general affinity for either single- or double-stranded DNA. Sequence-specific DNA-binding proteins generally interact with the major groove of B-DNA, because it exposes more functional groups that identify a base pair. However, there are some known narrow-groove DNA-binding ligands such as netropin, distamycin, etc.

DNA chip A high-density array of DNA molecules used for parallel hybridization analyses.

DNA computing The use of DNA and biotechnological methods to solve computational problems.

DNA cytosine methylase (Dcm) An enzyme involved in methylation of DNA in *E. coli*.

DNA Data Bank of Japan (DDBJ) A sequence database maintained by the Japan's National Institute of Genetics.

DNA deoxyribonucleic acid The chemical that forms the basis of the genetic material in virtually all organisms. DNA is composed of the four nitrogenous bases—adenine, cytosine, guanine, and thymine, which are covalently bonded to a backbone of deoxyribose phosphate group. Two complementary strands (where all Gs pair with Cs and As with Ts) form a double-helical structure which is held together by hydrogen bonding between the cognate bases and runs in opposite direction.

DNA-dependent DNA polymerase An enzyme that makes a DNA copy of a DNA template.

DNA-dependent RNA polymerase An enzyme that makes an RNA copy of a DNA template.

DNA fingerprinting A process which uses fragments of DNA to identify the unique genetic make-up of an individual. This technique is based on a restriction enzyme digest of tandemly repeated DNA sequences that are scattered throughout the human genome, but are unique to each individual.

DNA glycolyase An enzyme that cleaves the β-N-glycosidic bond between a base and the sugar component of a nucleotide as a part of the base excision and mismatch repair processes.

DNA gyrase A type II class of enzymes called topoisomerases, which function during DNA replication to relax positive supercoiling of the DNA molecule and which introduce negative supercoiling into non-supercoiled molecules early in the life cycle of many phages *E. coli*.

DNA ligase An enzyme that catalyses the formation of a co-valent bond between adjacent 5´-P and 3´-OH termini in a broken polynucleotide strand of double-stranded DNA.

DNA marker A DNA sequence that exists as two or more readily distinguished versions and which therefore can be used to mark a map position on a genetic map, physical map or integrated genome map.

DNA methylation The chemical modification of DNA by attachment of methyl groups, which has various regulatory functions in prokaryotes and eukaryotes.

DNA microarrays The deposition of oligonucleotides or cDNAs onto an inert substrate such as glass or silicon. Thousands of molecules may be organized spatially into a high-density matrix. These DNA chips may be probed to allow monitoring the expression of many thousands of genes simultaneously. Uses include study of polymorphisms in genes, de novo sequencing or molecular diagnosis of disease.

DNA photolyase A bacterial enzyme involved in photoreactivation repair.

DNA polymerase An enzyme that catalyses the synthesis of DNA from a DNA or RNA template. This enzyme assembles DNA into a double helix by adding complementary bases to a single strand of DNA. Linkages are formed by adding nucleotides at the 5´ hydroxyl group to the phosphate group located on the 3´ hydroxyl end.

DNA probe Short single-stranded DNA molecules of specific base sequence, labelled either radioactively or immuno-logically, that are used to detect and identify the complementary sequence in a gene or genome by hybridizing specifically to that gene or sequence.

DNA repair The biochemical processes that correct mutations arising from replication errors and the effects of mutagenic agents.

DNA repair gene A gene that encodes proteins that correct errors in DNA sequencing.

DNA replication The use of existing DNA as a template for the synthesis of new DNA strands.

DNase (Deoxyribonuclease) *See* deoxyribonuclease.

DNase I hypersensitive site A short region of eukaryotic DNA that is relatively easily cleaved with deoxyribonuclease I, possibly coinciding with positions where nucleosomes are absent.

DNA sequence The relative order of base pairs in DNA, it may be a DNA fragment, a gene, a chromosome, or an entire genome.

DNA sequencing A process of determining the order of nucleotides in a segment of DNA.

DNA shuffling A PCR-based procedure that results in directed evolution of a DNA sequence.

DNA topoisomerase An enzyme that introduces or removes turns from the double helix by breakage and reunion of one or both polynucleotides.

DNA transposon A transposon whose transposition mechanism does not involve an RNA intermediate.

DNA tumour virus A virus with a DNA genome, able to cause cancer after infection of a cell.

domain 1. A discrete portion of a protein assumed to fold independently of the rest of the protein and possessing its own function. It is a region of special biological interest within a single protein sequence. 2. The set of all allowable values that a database attribute may assume. A domain class is a group of domains that share a common set of well-defined properties or characteristics.

domain duplication Duplication of a gene segment coding for a structural domain in the protein product.

domain shuffling Rearrangement of segments of one or more genes, each segment coding for a structural domain in the gene product, to create a new gene.

dominant An allele that is expressed in a heterozygote.

dominant negative mutation The mutant polypeptide disrupts the function of the wild-type polypeptide in heterozygotes.

donor site The splice site at the 5´ end of an intron.

dosage compensation The imbalance caused by having two copies of the X chromosome in females compared to only one copy in males is countered (in humans) by X inactivation or (in *Drosophila*) by reducing the relative level of activity of X-linked genes in females.

dose repitition The presence of multiple copies of a DNA sequence which can be shown to produce increased quantities of a gene product relative to a single copy sequence.

dot matrix A method for nucleotide sequence alignment.

dot product The inner product of two vectors.

double helix The helical structure that is the natural form of DNA in the cell, where two linear DNA strands are bonded together by base pairing.

double heterozygote An individual who has two different gene mutations at two separate genetic loci.

double restriction Digestion of DNA with two restriction endonucleases at the same time.

double-strand break repair A model of recombination in which the exchange process is initiated by a duplex DNA molecule containing a double-stranded break.

double-stranded RNA (dsRNA) A structure formed when two complementary RNA sequences bind to each other. Transfer RNAs contain sections of dsRNA as do external guide sequences.

double-stranded RNA Adenosine Deaminase (dsRAD) An enzyme that edits various eukaryotic mRNAs by deaminating adenosine to inosine.

download To transfer a file from a remote host to a local machine via FTP.

downstream Towards the 3′ end of a string of nucleotides.

draft sequence The sequence generated by the HGP as of June 2000 that, while incomplete, offers a virtual road map to an estimated 95% of all human genes.

Drosophila melanogaster (D. melanogaster) A species of fruit fly used as a model organism to study genetics and development.

drug An agent that affects a biological process. Specifically, a molecule whose molecular structure can be correlated with its pharmacological activity.

drug discovery cycle The cycle of events required to develop a new drug. Typically this involves research, pre-clinical testing and a clinical development, and can take from 5 to 12 years.

Duchenne Muscular Dystrophy (DMD) A type of genetic disease that primarily affects voluntary muscles (enlargement of muscles).

duplex A double-stranded DNA or RNA, or a double helix formed by the complementary pairing of the nucleotides of a single-stranded DNA with an RNA molecule.

duplication The presence or the creation of two copies of a DNA segment in the genome.

DUST A program for filtering low complexity regions from nucleic acid sequences.

dynamical Changing over time.

Dynamic Allele-Specific Hybridization (DASH) This technique is based on differences in temperatures between duplexes resulting due to perfect match and mismatch between the PCR product and an oligonucleotide, 15–21 bases long.

dynamical system A system that changes over time according to a set of fixed rules that determine how one state of the system moves to another state.

dynamic programming A type of algorithm widely used for constructing sequence alignments and for evaluating all possible gene structures candidates.

E

EBI (European Bioinformatics Institute) An outstation of the European Molecular Biology Laboratory (EMBL). EBI is located in Hinxton, England, near Cambridge University.

EcoCyc A scientific database for the bacterium *Escherichia coli* K12 MG1655.

École Politechnique Fédérale de Lausanne (EPFL) The Swiss Federal Institute of Technology-Lausanne.

ecology The study of the relationships and interactions between organisms and then environment.

ecosystem A biological system consisting of many organisms from different species.

edge A connection between two vertices of a graph.

edge of chaos The hypothesis that many natural systems tend towards dynamical behaviour that borders static patterns and the chaotic regime.

edit distance The smallest number of insertions, deletions, and substitutions required to change one string or tree into another.

effector The part of a classifier system that can translate messages into actions that can manipulate a system or an environment.

EGCG Extensions to the GCG package.

Eidgenössische Technische Hochschule (ETH) Swiss Federal Institute of Technology, Zurich, Switzerland.

eigenvalue The change in length that occurs when the corresponding eigenvector is multiplied by its matrix.

eigenvector A unit length vector that retains its direction when multiplied with the matrix to which it corresponds.

electron density map The distribution of electron density in a crystal that is calculated from the X-ray diffraction pattern using a Fourier transform.

electronic Northerns The use of an electronic database of cDNA sequences (or probes derived from them) in order to measure the relative levels of mRNAs expressed in different cells or tissues. An example of the use of an electronic Northern might be to identify the differences in the genes expressed in prostate cancer and those in benign prostate hyperplasia, by subtracting the database of one from the other and seeing which cDNAs remain.

electrophoresis The primary method used to separate the mixture of nucleotide or peptide fragments generated from DNA or protein cleavage experiments. The apparatus consists of a plate coated with either agarose or acrylamide gels, which is placed in an electric field. As the solvent is allowed to infuse up the length of the plate, the components of the mixture are separated by size. In addition, the electrical charge along the side of the plate forces migration of the DNA or protein fragments according to the net charge of the residues.

electroporation A process using high-voltage current to make cell membranes permeable to allow the introduction of new DNA; commonly used in recombinant DNA technology.

electrostatic attraction The attraction between atoms of opposite charge that holds the atoms together in ionic bonds.

electrostatic interactions Ionic bonds that form between charged chemical groups.

electrostatic surface potential The electrostatic charges on the surface of a protein, often indicative of the protein's functional regions.

elongation The growth of the polypeptide chain through the addition of amino acids during the protein synthesis; the second step in translation.

elongation factor A protein that plays an ancillary role in the elongation step of transcription or translation.

EMACS A UNIX text editor.

embedding A method of taking a scalar time series and using delayed snapshots of the values at fixed time intervals in the past so that the dynamics of the underlying system can be observed as a function of the previously observed states.

EMBL The European Molecular Biology Laboratory in Heidelberg, Germany, maintains the EMBL database, one of the major public DNA sequence databases.

EMBL nucleotide sequence database Europe's primary nucleotide sequence resource. Main sources for DNA and RNA sequences are direct submissions from individual researchers, genome sequencing projects and patent applications. The database is produced in collaboration with GenBank and the DNA Database of Japan (DDBJ). Each of the three groups collects a portion of the total sequence data reported worldwide, and all new and updated database entries are exchanged between the groups on a daily basis.

embryo Term applied to the zygote after the beginning of mitosis that produces a multicellular structure.

Embryonic Stem cell (ES cell) A totipotent cell from the embryo of a mouse or other organism.

emergence The property of a collection of simple subunits that comes about through the interactions of the subunits, and is not a property of any single subunit.

end-labelling The attachment of a radioactive or other label to one end of a DNA or RNA molecule.

end-modification Chemical alteration of the end of an RNA molecule.

endocytosis The incorporation of materials from outside the cell by the formation of vesicles in the plasma membrane. The vesicles surround the material so that the cell can engulf it.

Endogenous Retrovirus (ERV) An active or inactive retroviral genome integrated into a host organism's genome.

endonuclease An enzyme that cleaves at internal locations within a nucleotide sequence. The enzyme's site of action is generally a sequence of 8 bases. For *E. coli*, treatment with a restriction endonuclease will lead to around 70 fragments. Cleavage of human DNA leads to around 50,000 fragments.

Endoplasmic Reticulum (ER) A network of membranous tubules in the cytoplasm of a cell; involved in the production of phospholipids, proteins, and other functions.

endosymbiont theory A theory that holds that the mitochondria and chloroplasts in eukaryotic cells are derived from symbiotic prokaryotes.

endosymbiosis Theory that attempts to explain the origin of the DNA-containing mitochondria and chloroplasts in early eukaryotes by the engulfing of various types of bacteria that were not digested but became permanent additions to the ancestral "eukaryote".

energy The ability to bring about changes or to do work.

energy flow The movement of energy through a community via feeding relationships.

enhancer DNA sequences that can greatly increase the transcription rates of genes even though they may be far upstream or downstream from the promoter they stimulate.

Ensembl A joint project between EMBL–EBI and the Sanger Centre (UK) to develop a software system which produces and maintains automatic annotation on eukaryotic genomes. Now the genomic data on humans is available.

entity A single object about which data can be stored.

Entity–Relationship Diagram (ERD) A specialized chart that illustrates the interrelationships between entities in a database.

Entrez An online retrieval system for searching several linked databases, provided by the National Center for Biotechnology Information (NCBI).

entropy (H) A measure of a system's degree of randomness or disorder.

environmental variance Within a population, the measure of the variation of a particular phenotype that is due to environmental factors, as opposed to variations in genotype.

enzyme A class of proteins that are capable of catalysing chemical reactions (the making or breaking of chemical bonds). They do so by orienting their substrates into a suitable geometry in a particular location (the active site) where electrophilic or nucleophilic amino acid residues can participate in the reaction. Enzymes are protein catalyst that speeds up chemical reactions that would otherwise be prohibitively slow under physiological conditions.

enzyme kinetics The investigation of the binding rate of an enzyme to its substrate.

EOF The ASCII character that tells the computer when the end of a computer file has been reached.

EOL The character that tells the computer when the end of a line of text has been reached.

epigenetic Any factor that influences the phenotype that is not part of the genotype.

epigenomics The study of complex expression networks or linkages both spatially (within the body) and temporally (at different times in development).

episome A plasmid that is able to integrate into the host cell's chromosome.

episome transfer Transfer of some or all of a bacterial chromosome by integrating it into a plasmid.

epistasis The masking of the effects of one gene by the action of another.

equilibrium A state in which a system will remain unchanged unless the system is subjected to some perturbation.

equilibrium constant Value that describes the equilibrium state of the reversible reaction between two molecular species.

ergodic The property of a dynamical system such that all regions of a state space are visited with similar frequency and that all regions will be revisited (within a small proximity) if enough time is given.

Escherichia coli (E. coli) A common bacterium that has been studied intensively by geneticists because of the small size of its genome, the organism's lack of pathogenicity, and its ease of growth in the laboratory.

E-site A position within a bacterial ribosome to which a tRNA moves immediately after deacylation.

EST (Expressed Sequence Tag) A partial sequence of a cDNA clone that can be used to identify sites in a gene. This sequence can be amplified by PCR. ESTs act as physical markers for cloning and full length sequencing of the cDNAs of expressed genes. Typically identified by purifying mRNAs, converting to cDNAs, and then sequencing a portion of the cDNAs.

ethidium bromide A type of intercalating agent that causes mutations by inserting between adjacent base pairs in a double-stranded DNA molecule.

Ethylmethane Sulphonate (EMS) A mutagen that acts by adding alkyl groups to nucleotide bases.

euchromatin The portions of a eukaryotic chromosome that are relatively uncondensed, and are thought to contain active genes.

euclidean Pertaining to a standard geometry.

eugenics The study of improving a species by artificial selection; usually refers to the selective breeding of humans.

eukaryote An organism whose DNA is enclosed in a membrane-bound nucleus.

Euler's method The simplest method to obtain a numerical solution of a differential equation.

euploid Containing a whole number of haploid sets of chromosomes.

European Bioinformatics Institute (EBI) *See* EBI.

European Molecular Biology Laboratory (EMBL) *See* EMBL.

European Molecular Biology Network (EMBnet) EMBnet was established in 1988, and provides services including local molecular databases and software for molecular biologists in Europe.

European Molecular Biology Open Software Suite (EMBOSS) A suite of freely available programs and libraries for molecular biology and bioinformatics.

E-value (Expectation value) The number of different alignments with scores equivalent to or better than S that are expected to occur in a database search by chance. The lower E-value, the more significant the score.

evolution A process operating on populations that involves variation among individuals, traits being inheritable, and a level of fitness for individuals that is a function of the possessed traits.

evolutionary game theory A game theory that describes the game models in which players choose their strategies through a trial- and-error process in which they learn over time that some strategies work better than others.

Evolutionary Stable Strategy (ESS) A strategy in game theory and biology, that, when possessed by an entire population, results in an equilibrium such that mutation of the strategy can never result in an improvement for an individual.

evolutionary tree A diagram showing the evolutionary history of organisms based on differences in amino acid sequences.

excision repair A DNA repair process that corrects various types of DNA damage by excising and resynthesizing a region of polynucleotide.

excitatory A neural synapse that is positive, so that activity at that connection encourages activity in the connected neuron.

exclusive OR (XOR) Given two Boolean inputs, the output of XOR is 1 if and only if the two inputs are different; otherwise, the output is 0.

executable file A file of commands (usually in machine language) that can be run on a computer.

exit site A position along a bacterial ribosome to which a tRNA molecule moves after deacylation.

exogenous DNA DNA originating outside an organism.

exon The region of DNA which encodes proteins. These regions are usually found scattered throughout a given strand of DNA. During transcription of DNA to RNA, the separate exons are joined to form a continuous coding region.

exon–intron boundary The nucleotide sequence at the junction between an exon and an intron.

exonuclease An enzyme which cleaves nucleotides sequentially starting at the free end of the linear chain of DNA.

exon skipping Aberrant splicing, in which one or more exons are omitted from the spliced RNA.

exon theory of genes An intron's early hypothesis that holds that introns were formed when the first DNA genomes were constructed.

exon trapping A method based on cloning, to identify the positions of exons in a DNA sequence.

experimentation A process by which one attempts to understand nature.

Expert Protein Analysis System (ExPASy) A set of databases, maintained by the Swiss Institute of Bioinformatics (SIB), dedicated to the analysis of protein sequences and structures.

expert system A computer program that contains expert knowledge about a particular problem, often in the form of a set of *if-then* rules.

exponential A function that is a constant times another constant to the power of the argument.

exportin A protein involved in transport of molecules out of the nucleus.

expression In relation to genes, the phenotypic manifestation of a trait. It is a measure of the presence, amount, and time-course of one or more gene products in a particular cell or tissue. Expression studies are typically performed at the RNA (mRNA) or protein level in order to determine the number, type, and level of genes that may be up-regulated or down-regulated during a cellular process, in response to an external stimulus, or in sickness or disease. Gene chips and proteomics now allow the study of expression profiles of sets of genes or even entire genomes.

expression profile The level and duration of expression of one or more genes, selected from a particular cell or tissue type, generally obtained by a variety of high-throughput methods, such as sample sequencing, serial analysis, or microarray-based detection.

expression vector A cloning vector that is engineered to allow the expression of protein from a cDNA. The expression vector provides an appropriate promoter and restriction sites that allow insertion of cDNA.

expressivity The degree of expression of a mutant phenotype.

extant organism An organism that currently exists.

Extensible Markup Language (XML) A universal format for structured documents and data on the web.

extensive form A graphical representation of a sequential game.

external node The end of a branch in a phylogenetic tree, representing one of the organisms or DNA sequences being studied.

extinction The elimination of all individuals in a group by natural (dinosaurs, trilobites) or human-induced (dodo, passenger pigeon) means.

extrachromosomal gene A gene in the genome of a mitochondrion or chloroplast.

extreme value distribution The probability distribution applicable to the scores of optimal local alignments.

F

facultative heterochromatin Chromatin that has a compact organization in some, but not all cells; thought to contain genes that are inactive in some cells or at some periods of the cell cycle.

familial trait A trait which is more common in the relatives of an affected person.

FASTA An alignment program for protein sequences created by Pearson and Lipman in 1988. The program is one of the many heuristic algorithms proposed to speed up sequence comparison. The basic idea is to add a fast pre-screen step to locate the highly matching segments between two sequences, and then extend these matching segments to local alignments using more rigorous algorithms such as Smith–Waterman.

FASTA format A sequence in FASTA format begins with a single-line description, followed by lines of sequence data. The description line is distinguished from the sequence data by a greater-than (">") symbol in the first column. It is recommended that all lines of text be shorter than 80 characters in length. An example for sequence in FASTA format is

```
>gi|532319|pir|
TVFV2E|TVFV2E envelope
protein        LRLRYCAPAG-
FALLKCND       ADYDGFKTNC-
SNVSVVHCTNLMNTT       VTT-
GLLLNGSYSENRTIWQKHR
TSNDALILLNKHYNLTVTCK
RPGNKT        VLPVTIMAGLVF-
HSQKYNLRLRQ        AWCHFP-
SNKGAWKEVKEEIVNLPK
ERYRGTNDPKRIFFQRQWGD.
```

fauna Term referring collectively to all animals in an area. The zoological counterpart of flora.

feature Annotation on a specific location of a given sequence.

fecundity The power of a species to multiply rapidly; its capacity to form reproductive elements.

federated database system A type of meta-database management system (DBMS) which transparently integrates multiple autonomous database systems into a single federated database. The constituent databases are interconnected via a computer network, and may be geographically decentralized. Since the constituent database systems remain autonomous, a federated database system is a contrastable alternative to the task of merging together several disparate databases. A federated database (or virtual database) is the fully integrated, logical composite of all constituent databases in a federated database system.

federated information systems Their main characteristic is that they are constructed as an integrating layer over existing legacy applications and databases. They can be broadly classified into three dimensions: the degree of autonomy—they allow in integrated components, the degree of heterogeneity between components they can cope with, and whether or not they support distribution. Whereas the communication and interoperation problem has come into a stage of applicable solutions over the past decade, semantic data integration has not become similarly clear.

feedback A loop in information flow or in cause and effect.

feedback neural network An artificial neural network in which every neuron is potentially connected to other neurons.

feedforward neural network An artificial neural network that is organized with separate layers of neurons.

Feigenbaum constant A constant number that characterizes when bump-like maps such as the logistic map will bifurcate.

FEN1 The "flap endonuclease" involved in replication of the lagging strand in eukaryotes.

fertility The number of live offspring per individual of a given genotype.

fertilization The fusion of two gametes (sperm and ovum) to produce a zygote that develops into a new individual with a genetic heritage derived from both parents.

field A single unit of data stored as part of a database record.

Field Inversion Gel Electrophoresis (FIGE) An electrophoresis method used to separate large DNA molecules.

File Transfer Protocol (FTP) An Internet computer protocol that allows electronic files to be sent and received in a uniform fashion across a computer network.

Filial generation (F_1, F_2) Each generation of offspring in a breeding program, designated as F_1, F_2, etc.

filtering The process of hiding regions of (nucleic acid or amino acid) sequence having characteristics that frequently lead to spurious high scores. It is also known as masking.

fingerprint A set of motifs used to predict the occurrence of similar motifs, in either an individual sequence or in a database. Fingerprints are refined by iterative scanning of a composite protein sequence database. A composite or multiple-motif fingerprint contains a number of aligned motifs taken from different parts of a multiple alignment. True family members are then easy to identify by virtue of possessing all elements of the fingerprint, while subfamily members may be identified by possessing only part of it.

fingerprinting The process of identifying overlapping regions at the ends of DNA fragments.

finished sequence High-quality, low-error, gap-free DNA sequence of the human genome.

finite state automaton (FSA) The simplest computing device. Although it is not nearly powerful enough to perform universal computation, it can recognize regular expressions. It is also known as finite state machine (FSM).

firewall A security barrier erected between an organization's internal network and a public computer network such as the Internet.

First law of thermodynamics Energy can neither be created nor destroyed; it changes from one form to another.

first normal form (1NF) A database relation in which each attribute is atomic, that is, it contains a single numeric or character-string value.

FISH (Fluorescence *in situ* hybridization) A method used to pinpoint the location of a DNA sequence on a chromosome.

fitness The ability of an organism or allele to survive and reproduce.

fitness landscape A representation of how mutations can change the fitness of one or more organisms.

fixation The situation when a single allele reaches a frequency of 100% in a population.

fixation probability The probability that a particular allele will achieve fixation in a population.

fixed point A point in a dynamical system's state space that maps back to itself.

flanking sequence Untranscribed sequences at the 5´ end or 3´ end terminal of transcribed genes.

flat file Pure text documents that are totally unstructured. This type of file generally does not provide very specific search answers, but it is the most popular type of file on the web and is now a bit easier to search, thanks to the use of hyperlinks.

flora Term collectively applied to all of the plants in an area. The botanical counterpart of fauna.

flow cytometry A method for the separation of chromosomes by detecting the light-absorbing or fluorescing properties of cells or subcellular fractions (i.e., chromosomes) passing in a narrow stream through a laser beam.

flow karyotyping Use of flow cytometry to analyse and separate chromosomes on the basis of their DNA content.

FLpter value The unit used in FISH to describe the position of a hybridization signal relative to the end of the short arm of the chromosome.

fluid-mosaic Widely accepted model of the plasma membrane in which proteins (the mosaic) are embedded in lipids (the fluid).

FlyBase A database of the *Drosophila* genome.

fold The overall folding pattern of a 3D protein structure.

foldback DNA DNA that contains a perfect or nearly perfect palindrome and is able to form a hairpin-like structure by folding back on itself when single-stranded.

folded leaf A layer of alpha-helices wrapped around a single hydrophobic core but not with the simple geometry of a bundle.

folding domain A domain, or segment, of a polypeptide that folds independently of the other segments.

food chain The simplest representation of energy flow in a community.

food pyramid A way of depicting energy flow in an ecosystem; shows producers (mostly plants or other phototrophs) on the first level and consumers on the higher levels.

food web A complex network of feeding interrelations among species in a natural ecosystem; more accurate and more complex depiction of energy flow than a food chain.

foreign key A field in a relational table that matches the primary key column of another table.

forensics The use of DNA for identification.

formal system A mathematical formalism in which statements can be constructed and manipulated with logical rules.

FORTRAN (Formula Translation) An early programming language designed for scientific computing.

fosmid A high-capacity vector carrying the F plasmid origin of replication and a lambda *cos* site.

founder effect The loss of genetic variation that occurs when a new population is established by a very small number of individuals from a larger population. As a result of the loss of genetic variation, the new population may be distinctively different, both genetically and phenotypically, from the parent population from which it is derived. In extreme cases, the founder effect is thought to lead to the speciation and subsequent evolution of new species.

four-fold degenerate site A nucleotide site within a codon at which all possible substitutions are synonymous, that is, they code the same amino acid.

fourth normal form (4NF) A normal form used in database normalization. Introduced by Ronald Fagin in 1977, 4NF is the next level of normalization after Boyce–Codd normal form (BCNF). Whereas the second, third, and Boyce–Codd normal forms are concerned with functional dependencies, 4NF is concerned with a more general type of dependency known as a multivalued dependency. A table is in 4NF if and only if, for each of its non-trivial multivalued dependencies $X \twoheadrightarrow Y$, X is a superkey—that is, X is either a candidate key or a superset thereof.

F plasmid A fertility plasmid that directs conjugal transfer of DNA between bacteria.

fractal A geometrical structure in which the same pattern is repeated at many different scales; a self-similar object.

fractal dimension An extension of the notion of dimension found in Euclidean geometry.

fragile site A position in a chromosome that is prone to breakage because it contains an expanded trinucleotide repeat sequence.

frameshift A deletion, substitution, or duplication of one or more bases that causes the reading frame of a structural gene to shift from the normal series of triplets.

frameshifted protein A protein encoded entirely or in part by a reading frame different from the original or main reading frame of a gene.

frameshifting The controlled movement of a ribosome from one reading frame to another at an internal position within a gene.

frameshift mutation A mutation resulting from insertion or deletion of a group of nucleotides that is not a multiple of three and which therefore changes the frame in which translation occurs.

fraternal twins Siblings born at the same time as the result of fertilization of two ova by two sperm.

free text In contradistinction to a controlled vocabulary, free text has no structured set of words, such that two related entries might not be identified in a search because different words are used to describe each entry.

full gene sequence The complete order of bases in a gene.

function A mapping from one space to another.

functional analysis Area of genome research devoted to identifying the functions of unknown genes.

functional bioinformatics The emerging field of functional bioinformatics focuses on the development of ontologies or concept classifications fed into algorithms used to perform computations of the functions of biomolecules.

functional constraint The degree of intolerance characteristic of a site or a locus towards nucleotide substitutions.

functional dependency A database property when one attribute in a relation uniquely determines another attribute.

functional domain A region of eukaryotic DNA around a gene or group of genes that can be delineated by treatment with deoxyribonuclease I. It is a well-defined region within a protein that can perform a specific function.

functional genomics The discovery and determination of the function of genes through systematic analysis of their activity in healthy and diseased tissues.

function approximation The task of finding an instance from a class of functions that is minimally different from an unknown function.

fusion protein Proteins that are created through the joining of two or more genes which originally coded for separate proteins. Translation of this fusion gene results in a single polypeptide with functional properties derived from each of the original proteins. Recombinant fusion proteins are created artificially by recombinant DNA technology for use in biological research or therapeutics. Chimeric mutant proteins occur naturally when a large-scale mutation, typically a chromosomal translocation, creates a novel-coding sequence containing parts of the coding sequences from two different genes. Naturally occuring fusion proteins are important in cancer, where they may function as oncoproteins.

fuzzy logic A superset of Boolean logic dealing with the concept of partial truth, in which numbers from 0 to 1 are used as truth values between "completely false" (0) and "completely true" (1).

G

gain-of-function mutation A mutation that confers some new function to an organism.

gamete A mature reproductive cell (sperm or ovum) that fuses with a second gamete to produce a new cell during sexual reproduction.

game theory A mathematical formalism used to study human games, economics, military conflicts and biology.

gap A space introduced into an alignment to compensate for insertions and deletions in one sequence relative to another. To prevent the accumulation of too many gaps in an alignment, introduction of a gap causes the deduction of a fixed amount (the gap score) from the alignment score. Extension of the gap to encompass additional nucleotides or amino acid is also penalized in the scoring of an alignment.

gap gene A gene that aids in establishing positional information within the developing embryo.

gap opening penalty The penalty in an alignment score for a gap of any length.

gapped alignment An alignment in which gaps are permitted.

gap penalty An assessment of how frequently gap events occur in evolution compared to the frequency of point mutations.

gap period One of the two intermediate periods within the cell cycle.

gap score The score assigned to a gap.

gcc A "C" compiler produced by the GNU Free Software Foundation.

GC content The percentage of nucleotides in a genome that are guanine (G) or cytosine (C).

GCG assembly A tool using the GCG Fragment Assembly System created by Genetics Computer Group, Inc.

GC-rich area Many DNA sequences carry long stretches of repeated G and C which often indicate a gene-rich region.

gel A dense network of fine particles dispersed in water.

gel electrophoresis A method of separating large molecules (such as DNA fragments or proteins) from a mixture of similar molecules. An electric current is passed through a medium containing the mixture, and each kind of molecule travels through the medium at a different rate, depending on its electrical charge and size.

gel retardation analysis A technique that identifies protein-binding sites on DNA molecules by virtue of the effect that a bound protein has on the mobility of the DNA fragments during gel electrophoresis.

gel stretching A technique for preparing restricted DNA molecules for optical mapping.

GenBank The public DNA sequence database maintained by the National Center for Biotechnology Information (NCBI). This databank of genetic sequences is operated by a division of the National Institutes of Health.

GenBank Identifier (GI) A unique number assigned to protein and nucleotide sequences in the GenBank database.

gene The fundamental physical and functional unit of heredity, i.e., a section of DNA at a specific position on a particular chromosome that specifies the amino acid sequence for a protein.

gene arrays The covalent attachment of oligonucleotides or cDNA directly onto a small glass or silicon chip in organized arrays. Over 50,000 different DNA fragments can be presented on a single chip providing a high-throughput parallel method of probing gene expression, genotype or gene function.

gene chip technology Development of cDNA microarrays from a large number of genes.

gene cloning Isolating a gene and producing many identical copies of it.

gene conversion A process that results in the four haploid products of meiosis, displaying an unusual segregation pattern.

gene diversity A measure of genetic variability in a population: the mean expected heterozygosity per locus in a population.

gene dosage The number of copies of a gene which are present in the genome.

gene expression The conversion of information from gene to protein via transcription and translation.

Gene Expression Markup Language (GEML) An openstandard XML format for DNA microarray and gene expression data.

gene family Group of closely related genes that make similar products, i.e., two or more genes are related by divergent evolution from a common ancestor, either by speciation or gene duplication.

gene flow The movement of alleles from one population into another. It is also called as gene migration.

gene fragment A DNA fragment consisting of short isolated regions from within a gene.

gene index A listing of the number, type, label and sequence of all the genes identified within the genome of a given organism. Gene indices are usually created by assembling the overlapping EST sequences into clusters, and then determining if each cluster corresponds to a unique gene. It is a method by which a cluster can be identified as representing a unique gene includes identification of long open reading frames (ORFs), comparison to genomic sequence, and detection of SNPs or other features in the cluster that are known to exist in the gene.

gene library A collection of cloned DNA fragments created by restriction endonuclease digestion that represent part or all of an organism's genome.

gene map A diagram of the relative positions of genes within the genome.

gene mapping Determination of the relative positions of genes on a DNA molecule (chromosome or plasmid) and of the distance, in linkage units or physical units, between them.

Gene Ontology (GO) "The goal of the Gene Ontology Consortium is to produce a dynamic controlled vocabulary that can be applied to all eukaryotes even as knowledge of gene and protein roles in cells is accumulating and changing."

gene pool All the genes in a sexually reproducing population, i.e., the complete set of unique alleles in a species or population.

gene prediction Predictions of possible genes made by a computer program based on how well a stretch of DNA sequence matches known gene sequences.

gene product The product, either RNA or protein, that results from expression of a gene. The amount of gene product reflects the activity of the gene.

gene profiling The expression pattern of a gene.

General Pathway Simulator (GEPASI) Software package used for modelling the biochemical systems.

general recombination Recombination between two homologous double-stranded DNA molecules.

General transcription factor (GTF) A protein or protein complex that is a transient or permanent component of the initiation complex formed during eukaryotic transcription.

generation time The average time span between two successive generations, i.e., the mean age of the parents at which they give birth to their middle child.

gene silencing Process which deactivates harmful foreign genes and for normal growth and development of plants; it can occur at either the transcriptional or post-transcriptional levels.

gene substitution The replacement of an allele—that at one time was fixed in the population—by a second allele, that arose by mutation and increased in frequency until it reached fixation.

gene superfamily A group of two or more evolutionarily related multigene families.

gene therapy A set of techniques used to treat inherited diseases. A healthy gene is inserted into the cells of a patient's body to counteract or overcome the effects of the defective gene.

Genetic Algorithm (GA) An (Artificial Intelligence) AI technique that solves problems by simulating the "survival of the fittest" among possible solutions.

genetic code The rules for converting the sequence of nucleotides in mRNA into a sequence of amino acids in a polypeptide.

genetic counselling Provides education and information about genetic-related conditions to patients and their families and helps them make informed decisions.

genetic discrimination Prejudice against those who have or are likely to develop an inherited disorder.

genetic disease A disorder, which may or may not be apparent at birth, which is a consequence of a mutation present in one or more of the patient's genes.

genetic divergence The separation of a population's gene pool from the other population due to mutation, genetic drift, and selection.

genetic engineering The technology used to genetically manipulate the living cells and to produce new chemicals or perform new functions.

genetic footprinting A technique for the rapid functional analysis of many genes at once.

genetic heterogeneity Similar phenotypes caused by mutations in more than one gene.

genetic illness Sickness, physical disability, or other disorder resulting from the inheritance of one or more deleterious alleles.

genetic linkage The physical association between two genes that are on the same chromosome.

genetic linkage data A genetic linkage map of the chromosome indicating the relative locations of known genes within the DNA sequence of the chromosome.

genetic map (linkage map) A physical map of a gene containing locations of known segments of DNA sequence with identifiable functions.

genetic mapping A technique to identify regions of the genome that may contain a gene associated with a particular disease. Genetic mapping requires sample—collection of DNA from families affected by the disease.

genetic marker A segment of DNA with an identifiable physical location on a chromosome, i.e., any gene that can be readily recognized by its phenotypic effect, and which can be used as a marker for a cell, chromosome, or individual carrying that gene. Also, any detectable polymorphism used to identify a specific gene.

genetic mutation An inheritable alteration in DNA or RNA which results in a change in the structure, sequence, or function of a gene.

genetic polymorphism The occurrence of one or more different alleles at the same locus in one per cent or greater of a specific population.

genetic predisposition Susceptibility to a genetic disease which may or may not result in actual development of the disease.

genetic profile The particular arrangement of genes and markers in the DNA, unique to each individual.

Genetic Programming (GP) A method of applying simulated evolution on programs or program fragments.

genetic redundancy The situation in which two genes in the same genome perform the same function.

genetics The study of the patterns of inheritance of specific traits.

genetic testing A technique used to determine whether an individual has a disease-causing allele of a certain gene. This test identifies individuals at risk for a specific genetic disease or at risk for transmitting such a disease to their children.

genetic variance Within a population, the measure of how much of the variation of a particular phenotype is due to genotypic variation, as opposed to environmental factors.

gene transfer Incorporation of new DNA into an organism's cells, usually by a vector such as a modified virus.

gene tree A phylogenetic tree that shows the evolutionary relationships between a group of genes.

gene-within-a-gene A gene whose intron contains a second gene.

genome The complete genetic content of an organism.

genome annotation It is now apparent that the bottleneck in genomics is no longer in sequencing the genomes, but lies in their annotation. Large-scale annotation efforts require handling massive amounts of genome data through automated pipelines, with a need to combine diverse sources of data and methods. In addition, it requires visualization tools to manually examine the automatic annotation, since integration of human expertise to assess the validity and authenticity of all computational results goes a long way to improve the quality of gene annotation.

genome project Research and technology development effort aimed at mapping and sequencing some or all of the genome of human beings and other organisms.

Genomes To Life (GTL) A research program of the DOE, planned as a successor to the human genome project, that seeks to use genomic data to understand fundamental biological processes.

Genome Survey Sequences (GSS) This DDBJ/EMBL/Gen-Bank division is similar in nature to the EST division, except that its sequences are genomic in origin, rather than cDNA.

genome-wide repeat Sequences that recur at many dispersed positions within a genome.

genomic compartmentalization The existence of independently replicated genomes in a cell. Usually, in reference to the genomes of organelles such as mitochondria and chloroplasts.

genomic DNA DNA derived from the genome, as distinct to cDNA, i.e., DNA sequence typically obtained from mammalian or other higher-order species, which includes both intron and exon sequence (coding sequence), as well as noncoding regulatory sequences such as promoter, and enhancer sequences.

genomic imprinting Inactivation by methylation of a gene on one of a pair of homologous chromosomes.

genomic library A collection of clones made from a set of randomly generated overlapping DNA fragments representing the entire genome of an organism.

genomics Sequencing and characterizing of the genome and analysis of the relationship between gene activity and cell function.

genomic sequence The order of the subunits, called bases, that make up a particular fragment of DNA in a genome.

genotype A description of the genetic composition of an individual, including the alleles that do not show any outward characteristics.

genotyping The use of markers to organize the genetic information found in individual DNA samples and to measure the variation between such samples.

genus Taxonomic subcategory within a family, composed of one or more species.

geometric mean The nth root of the product of n terms.

germ line The continuation of a set of genetic information from one generation to the next.

germ-line gene therapy An experimental process of inserting genes into germ cells or fertilized eggs to cause a genetic change that can be passed on to offspring.

germ-line mosaicism Two population of cells making up the germ-line of one individual.

Giemsa banding The pattern of dark- and light-coloured Giemsa-stained bands on metaphase chromosomes.

glider gun An object in Conway's Game of Life that builds and emits gliders, which can then be collided in purposeful ways to construct more complicated objects.

global alignment The alignment of two nucleic acid or protein sequences over their entire length.

global maximum The point in a search space that attains the highest possible value.

glutamic acid (Glu, E) One of the 20 amino acids and its codons are GAA and GAG. It is a non-essential amino acid.

glutamine (Gln, Q) One of the 20 amino acids encoded by the standard genetic code. Its codons are CAA and CAG.

glutamine-rich domain A type of activation domain.

glycine (Gly, G) An organic compound with the formula NH_2CH_2COOH. It is the smallest of the 20 amino acids commonly found in proteins, coded by codons GGU, GGC, GGA and GGG.

glycobiology The study of glycosylation—the addition of a complex series of sugar residues to proteins after translation of protein from an mRNA.

glycosylase An enzyme that cleaves the beta-N-glycosidic bond between a base and the sugar component of a nucleotide, as part of the base excision and mismatch repair process.

glycosylation The attachment of sugar units to a polypeptide, i.e., the addition of carbohydrate groups (sugars), e.g. to polypeptide chains.

Golgi complex An organelle in the animal cells composed of a series of flattened sacs that sort, chemically modify, and package proteins produced on the rough endoplasmic reticulum.

G_0 phase A cell may pause in the G_1 phase before entering the S phase and enter a state of dormancy called the G_0 phase. Most mammallian cells do this. In order to divide, the cell re-enters the cycle in S phase.

G₁ phase The first gap period in the cell cycle during interphase, after cytokinesis and before the S phase. For many cells, this phase is the major period of cell growth during its lifespan. During this stage, new organelles are being synthesized, so the cell requires both structural proteins and enzymes, resulting in great amount of protein synthesis. The metabolic rate of the cell will be high.

G₂ phase The third, final and usually the shortest sub-phase during interphase within the cell cycle in which the cell undergoes a period of rapid growth to prepare for mitosis. The G₂ phase prepares the cell for mitosis (M phase) which is initiated by prophase.

gradient The slope of a high-dimensional surface.

Grail A program for predicting coding regions and intron–exon boundaries of genomic sequences.

graph A set of points (vertices) and the connections (edges) between them.

Graphical User Interface (GUI) The part of a computer application by which the user interacts using graphical metaphors, instead of typed commands.

GRASP Graphical Representation and Analysis of Structural Properties. A molecular visualization and analysis program. It is particularly useful for the display and manipulation of the surfaces of molecules and their electrostatic properties.

greedy algorithm An algorithm that always takes the best immediate (or local) solution while finding an answer.

greek-key A topology for a small number of beta-sheet strands in which some inter-strand connections go across the end of barrel or, in a sandwich fold, between beta-sheets.

grep A UNIX command that lets you search for the presence of a string within a file.

Group I intron Large self-splicing ribozymes. They catalyse their own excision from mRNA, tRNA and rRNA precursors in a wide range of organisms. The secondary structure mark-up for this family represents only this conserved core. Group I catalytic introns often have long open reading frames inserted in loop regions.

Group II intron A class of self-catalytic ribozymes and retroelements found in rRNA, tRNA, mRNA of organelles in fungi, plants, protists, and bacteria. In contrast to group I introns, intron excision occurs in the absence of GTP and involves the formation of a lariat, with a branchpoint strongly resembling that found in lariats formed during splicing of nuclear pre-mRNA.

Group III intron A class of introns found in mRNA genes of chloroplasts in euglenoid protists. They have a conventional group II-type dVI with a bulged adenosine, a streamlined dI, no dII-dV, and a relaxed splice site consensus. Splicing is by two transesterification reactions with a dVI bulged adenosine as initiating nucleophile; the intron is excised as a lariat.

GTPase Activating Protein (GAP) A protein intermediate in the Ras signal transduction pathway.

GU–AG intron The commonest type of intron in eukaryotic nuclear genes.

guanine (G) A nitrogenous purine base, one of the molecular components of DNA and RNA. Always bonds with cytosine (G–C).

guanine methyltransferase An enzyme that attaches a methyl group to the 5´ end of a eukaryotic mRNA during the capping reaction.

Guanine Nucleotide-Releasing Protein (GNRP) A set of proteins that are intermediates in the Ras signal transduction pathway.

guanylyl transferase An enzyme that attaches a GTP to the 5´ end of a eukaryotic mRNA at the begining of the capping reaction.

guide RNA A short RNA that specifies the positions at which one or more nucleotides are inserted into an abbreviated RNA by pan-editing.

H

H H is the relative entropy of the target and background residue frequencies. H can be thought of as a measure of the average information (in bits) available per position that distinguishes an alignment from chance. At high values of H, short alignments can be distinguished by chance, whereas at lower H values, a longer alignment may be necessary.

haemoglobin The primary functional protein of red blood cells.

hairpin A double-helical region in a single DNA or RNA strand formed by the hydrogen-bonding between adjacent inverse complementary sequences to form a hairpin-shaped structure.

halophile A group of archaea that are able to tolerate high salt concentrations.

halting problem The problem of determining if a complete program will or will not halt on a particular input.

Hamiltonian path A cycle in a directed or undirected graph such that every vertex of the graph is in the cycle, but each vertex appears exactly once (except the first and last vertices).

Hamming distance The number of bits which differ between two binary strings of equal length.

haploid A nucleus (or cell) that has a single copy of each chromosome, half of the full set of genetic material.

haploid set The chromosomes in a haploid cell or individual.

haploinsufficiency The situation where inactivation of a gene on one of a pair of homologous chromosomes results in a change in the phenotype of the mutant organism.

haplotype A collection of alleles that are usually inherited together.

hardware Computer system made up of all physical components which are distinguished from the programs and data that are manipulated by the computer.

Hardy–Weinberg equilibrium A condition under which genotype frequencies in a diploid population are equal to the products of their allele frequencies.

Hardy–Weinberg law A rule which relates the frequencies of genotypes at a locus in a population to the frequencies of the alleles at that locus.

hash table A listing of the positions at which all words of a fixed length (word size or ktup) occurs in a query sequence.

Health Insurance Portability and Accountability Act of 1996 (HIPAA) A U.S. Federal law that requires the protection of information about patients and research subjects, and has broad impact on all clinical research activities.

helical wheel A graphical view of the asymmetry of the distribution of hydrophobic side chains in potential alpha-helices.

helicase An enzyme that breaks base pairs in a double-stranded DNA molecule.

Helix-Turn-Helix Motif (HTH) A common structural motif for binding of a protein to DNA that consists of two alpha-helices connected by a short non-helical segment called a "turn".

hemizygous Having only one copy of a chromosome rather than the usual two.

Hénon map A chaotic system that has a fractal strange attractor and operates in discrete time.

hereditary cancer Cancer that occurs due to the inheritance of an altered gene within a family.

heritability The extent to which a trait is genetically determined.

heterochromatin Chromatin that is relatively condensed and is thought to contain DNA that is not being transcribed.

heterodimer Protein composed of two different chains or subunits.

heteroduplex Hybrid structure formed by the annealing of two DNA strands (or an RNA and DNA) that have sufficient complementarity in their sequence to allow hydrogen bonding.

heteroduplex analysis Transcript mapping by analysis of DNA–RNA hybrids with a single-strand-specific nuclease such as S1.

heterogametic sex The sex with two different chromosomes, e.g. humans (males) and *Drosophila*.

heterogeneous nuclear RNA (hnRNA) RNA transcripts in the nucleus, representing precursors and processing intermediates of rRNA, mRNA, and tRNA, as well as mature RNA transcripts not yet transported into the cytoplasm.

heterokaryon A cell with more than one nucleus formed by the fusion of two or more cells (usually of different species).

heteromorphism Normal variation in the shape or staining of a chromosome.

heteropolymer An artificial RNA comprising a mixture of different nucleotides.

heterozygosity A measure of genetic variation in a population calculated either as the mean frequency of heterozygotes over all loci (observed heterozygosity), or as the mean frequency of heterozygotes expected in a population in Hardy–Weinberg equilibrium (expected heterozygosity or gene diversity).

heterozygote An individual diploid with different alleles at the locus.

heterozygous A diploid nucleus or cell that contains two different alleles for a particular gene.

heuristic algorithm An economical strategy for deriving a solution to a problem for which an exact solution is computationally impractical or intractable.

heuristics "Guesses" made by a program to obtain approximately accurate results, typically based on computationally tractable "rule of thumb" methods, rather than formal approaches. It describes how things are commonly understood, without resorting to deeper or more formal knowledge.

HGSI Human Genome Sequencing Index. A service provided by the NCBI to members of the international consortium to support coordination and tracking of the Human Genome Project (HGP). Sequence and mapping target data from centres participating in the international consortium are submitted via the HGSI website. This website also presents an overview of HGP progress to the research community in tabular and graphical displays of the target data.

hidden layer In a feed-forward or recurrent neural network, a layer of neurons that lie between the input layer and the output layer.

Hidden Markov Model (HMM) A computer algorithm which locates the essential, unique features which can distinguish a protein or gene family by analysing a range of known sequences from the family. Then these features are used to locate similar characteristics in unknown sequences.

higher taxon A taxon above the species level.

highly conserved sequence DNA sequence that is very similar across several different types of organisms.

highly repetitive DNA The fraction of genomic DNA consisting of sequences repeated on the average hundreds of thousands of times.

highly repetitive genes Functional genes appearing in numerous copies in the haploid genome.

High-Performance Liquid Chromatography (HPLC) A method for characterizing proteins. A method for charac-

terizing proteins, i.e., a form of column chromatography used frequently in biochemistry and analytical chemistry to separate, identify, and quantify the compounds. HPLC utilizes a column that holds chromatographic packing material (stationary phase), a pump that moves the mobile phase(s) through the column, and a detector that shows the retention times of the molecules. Retention time varies depending on the interactions between the stationary phase, the molecules being analysed, and the solvent(s) used.

High-Scoring Segment Pair (HSP) Local alignments with no gaps that achieve one of the highest alignment scores in a given search.

high-throughput bioinformatics Bioinformatics is currently undergoing dramatic changes, as high-throughput laboratory methods lead to changes in key approaches, including sequence analysis, gene expression analysis, protein expression analysis, and protein structure prediction and modelling.

high-throughput DNA sequencing Experimental procedures for determining massive amounts of genomic DNA or cDNA sequence data using highly automated sequencing machines.

High-Throughput Genome Sequence (HTGS, HTG) Unfinished genome sequence data; often with large numbers of gaps in the nucleotides, low accuracy, and lack of annotations.

high-throughput screening The method by which very large numbers of compounds are screened against a putative drug target in either cell-free or whole-cell assays. Typically, these screenings are carried out in 96 well plates using automated, robotic station-based technologies or in higher-density array ("chip") formats.

hill-climbing One of the simplest search methods that attempt to find a local maximum by moving in an uphill direction.

Histidine (His, H) One of the 20 standard amino acids present in proteins. In the nutritional sense, in humans, histidine is considered as an essential amino acid, but mostly only in children. Its codons are CAU and CAC.

histone One of the basic proteins found in nucleosomes and also the basic building blocks of chromatin.

HLA complex Another name for the MHC in humans. It refers to the "Human Leukocyte Antigen" complex located on chromosome 6, i.e., antigens in the region of chromosome 6 containing genes, that code for proteins that enable the immune system to differentiate tissues or proteins, between "self " and "non-self." These loci are identified by numbers and letters, such as HLA-B27.

HMMer HMMer is a freely distributable implementation of profile HMM software for protein sequence analysis.

Holliday structure An intermediate structure formed during recombination between two DNA molecules.

holoenzyme The version of the *E. coli* RNA polymerase, subunit composition $\alpha 2\beta\beta$'s, that is able to recognize promoter sequences.

homeobox A DNA-binding motif found in many proteins involved in developmental regulation of gene expression, i.e., a highly conserved region in a homeotic gene composed of 180 bases (60 amino acids) that specifies a protein domain (the homeodomain) that serves as a master genetic regulatory element in cell differentiation during development in species as diverse as worms, fruit flies, and humans.

homeodomain A 60-amino-acid protein domain coded

for by the homeobox region of a homeotic gene.

homeotic gene A gene that controls the activity of other genes involved in the development of a body plan. Homeotic genes have been found in organisms ranging from plants to humans.

homeotic mutation A mutation that results in the transformation of one body part into another.

homeotic selector gene A gene that establishes the identity of a body part such as a segment of the *Drosophila* embryo.

home page A document on the World Wide Web that acts as a front page or point of welcome to a collection of documents that may introduce an individual, organization, or point of interest.

homoduplex A double-stranded DNA, the complementary strands of which are derived from the same individual.

homolog One of a pair of chromosomes, in which one is obtained from the organism's maternal parent and the other from the paternal parent. It is found in diploid cells.

homologous In phylogenetics, describing particular features in different individuals that are genetically descended from the same feature in a common ancestor. In molecular biology, often "homologous" simply means similar, regardless of genetic relationship.

homologous chromosome A chromosome containing the same linear gene sequences as another, each derived from one parent.

homologous chromosomes Two or more identical chromosomes present in a single nucleus.

homologous genes Genes that share a common evolutionary ancestor.

homologous protein A protein that is related to another by common evolutionary history.

homologous recombination Recombination between two homologous double-stranded DNA molecules, that is, ones which share extensive nucleotide sequence similarity.

homology Similarity in DNA or protein sequences attributed to descent from a common ancestor. Two or more gene or protein sequences that share a significant degree of similarity, typically measured by the amount of identity (in the case of DNA), or conservative replacements (in the case of protein), that they register along their lengths. Sequence "homology" searches are typically performed with a query DNA or protein sequence to identify known genes or gene products that share significant similarity and hence might inform on the ancestry, heritage and possible function of the query gene.

homology domain A region in a protein sequence with similarity to an otherwise unrelated protein. This term should be used only if the region is of a size sufficient to form a domain.

homology modelling The use of 3-dimensional (3D) geometry and sequence information from proteins of known 3D structure to develop models for proteins whose 3D structure is unknown. In the first step of homology modelling, search and alignment algorithms are used to find the best sequence overlap of the "unknown" protein with the sequences of related proteins which have 3D data. In the second step, the geometry of the 3D structures are used as a template for generating a 3D structural model for the regions of high sequence homology in the unknown protein (the conserved regions). Finally, the sections with low homology to known proteins (the variable regions) are modelled using a variety of computational techniques.

homology searching A search for genes with sequences similar to that of an unknown gene in order to gain an insight into the function of the unknown gene.

homoplasy Similarity that has evolved independently and does not indicate a common phylogenetic origin.

homopolymer An artificial RNA comprising just one nucleotide.

homozygote A diploid individual with identical alleles at one or more loci.

homozygous A diploid nucleus that contains two identical alleles for a particular gene.

hopfield network A type of feedback neural network that is often used as an associative memory or as a solution to a combinatorial optimization problem.

horizontal gene transfer Any process in which an organism incorporates genetic material from another organism without being the offspring of that organism. Also known as lateral gene transfer (LGT). By contrast, vertical gene transfer occurs when an organism receives genetic material from its ancestor, e.g. its parent or a species from which it evolved.

Most thinking in genetics has focused upon vertical gene transfer, but there is a growing awareness that horizontal gene transfer is a highly significant phenomenon, and amongst single-celled organisms perhaps the dominant form of genetic transfer. Artificial horizontal gene transfer is a form of genetic engineering.

hormone response element A nucleotide sequence upstream of a gene that mediates the regulatory effect of a steroid hormone.

host Any computer on the Internet that can be addressed directly through a unique IP address.

hotspot A segment of genomic DNA that shows a high propensity to undergo mutation either spontaneously or under the action of a particular mutagen.

housekeeping gene A gene which is expressed at a similar level in almost all cells, presumably because its product is required for cell viability.

Hsp70 chaperone A family of proteins that bind to hydrophobic regions in other proteins in order to aid their folding.

HSSP Database of homology-derived structures of proteins.

Human and Molecular Genetics Center (HMGC) A centre of excellence in functional genomics, located at the Medical College of Wisconsin, Milwaukee, Wisconsin.

Human Anti-Murine Antibody Response (HAMA) An immune response generated in humans to antibodies raised in murine (e.g. mouse or rat) cells.

Human–Computer Interaction (HCI, CHI) The study of the psychology and design principles associated with the way humans interact with computer systems.

human–computer interface The "view" presented by a program to its user.

human gene therapy Insertion of normal DNA directly into cells to correct a genetic defect.

Human Genome Initiative (HGI) Collective name for several projects begun in 1986 by DOE to create an ordered set of DNA segments from known chromosomal locations, to develop new computational methods for analysing genetic map and DNA sequence data, and develop new techniques and instruments for detecting and analysing DNA.

Human Genome Mapping Project (HGMP) The HGMP

resource centre is an academic institution that provides a number of services, including access to databases, mirrors of databases, and access to extensive services and software for registered academic users.

Human Genome Project (HGP) A world-wide project aimed at deciphering all the three billion bases of the human genome, including mapping and sequencing every gene. This information will help to more rapidly identify genes causing diseases in humans.

Human Immunodeficiency Virus (HIV) The retrovirus that attacks T cells in the human immune system, destroying the body's defences and allowing the development of AIDS.

Huntington disease A progressive and fatal disorder of the nervous system that develops between the ages of 30 and 50 years. It is caused by an expansion of a trinucleotide repeat and inherited as a dominant trait.

hybrid An organism made by crossing two different species.

hybrid dysgenesis A syndrome of correlated abnormalities, that is, spontaneously induced in one type of hybrid between certain mutually interactive strains of *Drosophila*, but not in the reciprocal hybrid.

hybrides (or hybride molecular complexes) The formation of a complimentary complex between a probe molecule and a target molecule. This complex is generally tagged with a radioactive label on the probe molecule, so that the complex can be located and isolated for further study. Hybrid molecular complexes of the type DNA–DNA, DNA–RNA, and protein–protein are frequently used in genetic analysis. Since hybridization reactions are specific, they can be used to locate one DNA, RNA, or protein molecule within complex mixtures of similar molecules.

hybridization The process of joining two complementary strands of DNA or each one of DNA and RNA to form a double-stranded molecule. Hybridization for nucleic acid analysis requires that a probe of complementary sequence bind to a target sequence within the sample DNA.

hybridization array technologies The technology and manufacture of DNA or biochips and related microarray assays.

hybridization probing A technique that uses a labelled nucleic acid molecule as a probe to identify the complementary or homologous molecule to which it forms base pairs.

hydrogen bond A weak chemical interaction between an electronegative atom (e.g. nitrogen or oxygen) and a hydrogen atom that is covalently attached to another atom. This bond maintains the two-helices of DNA together and is also the primary interaction between water molecules, i.e., a dipole–dipole attraction in which a hydrogen atom bridges two electronegative atoms. One half of the hydrogen bond is a covalent bond and the other is an electrostatic bond.

hydrophilic An amino acid residue with a polar side chain, such as lysine or aspartic acid, that can form a hydrogen bond with water.

hydrophilicity The degree to which a molecule is soluble in water. Hydrophilicity depends to a large degree on the charge and polarizability of the molecule and its ability to form transient hydrogen bonds with (polar) water molecules.

hydrophobic An amino acid residue with an aliphatic or aromatic side chain, such as phenylalanine, that cannot bond with water.

hydrophobic effect A chemical interaction that results in hydrophobic groups becoming buried inside a protein.

hydrophobicity The degree to which a molecule is insoluble in water, and hence is soluble in lipids. If a molecule lacking polar groups is placed in water, it will be entropically driven to finding a hydrophobic environment (such as the interior of a protein or a membrane).

hydrophobic moment A quantitative measure of the asymmetry of the distribution of hydrophobic side chains in alpha-helices and beta-sheets.

hydrosphere The part of the physical environment that consists of all the liquid and solid water at or near the earth's surface.

hyperlink A connection between hypertext documents that allows a reader to trace concepts appearing in one document to related occurrence in other documents.

hypermedia A method of presenting interlinked multimedia information, such as texts, images, video, audio, and software applications.

hypermutation An increase in the mutation rate of a genome.

hypertext A method of presenting documents that allows them to be read in a richly interconnected way.

Hypertext Markup Language (HTML) A standard, text-based language used to create hypertext documents to be viewed on the World Wide Web.

Hypertext Transfer Protocol (HTTP) Communication protocol used on the Internet for the transfer of HTML documents.

hypervariable site A DNA or protein region that exhibits excessive intraspecific variability. Maintenance of so much variability usually requires the locus to be subject to a form of balancing selection, such as overdominance.

hypothesis An idea that can be experimentally tested; an idea with the lowest level of confidence.

I

ice age Interval of geological time between 2 million and 10,000 years ago during which the northern hemisphere experienced several episodes of continental glacial advance and retreat along with a climatic cooling.

identical twin Twins produced by the division of a single zygote; both have identical genotypes.

identity The extent to which two nucleotides or amino acid sequences are invariant.

identity matrix A scoring matrix in which only identical characters receive a positive score; the matrix has ones along the main diagonal and zeroes elsewhere.

idiopathic Arising spontaneously or from an obscure or unknown cause.

idiotype Antibody variants localized to the variable portion of an immunoglobulin that are recognized by their antigenic determinants. The determinants are composed from the antigen-combining site or CDRs. Every unique antigenic determinant has a specific antibody with its own unique idiotype.

imaginary number The square root of a negative number; a real number multiplied by the imaginary unit.

immunocytochemistry A technique that uses antibody probing to locate the position of a protein in a tissue.

immunoelectron microscopy An electron microscopy technique that uses antibody labelling to identify the positions of specific proteins on the surface of a structure such as a ribosome.

immunoglobulin A member of the globulin protein family consisting of two light and two heavy chains linked by disulphide bonds. All antibodies are immunoglobulins.

immunoscreening The use of an antibody probe to detect a polypeptide synthesized by a cloned gene.

immunotherapy Using the immune system to treat disease, for example, in the development of vaccines.

implication The Boolean "implies" ("if A then B") function.

implicit parallelism An idea that genetic algorithms have an extra built-in form of parallelism that is expressed when a GA searches through a search space.

importin A protein involved in transport of molecules into the nucleus.

imprinting The differential expression of genes depending on whether they were inherited maternally or paternally.

inborn error of metabolism A mutation that results in the absence of an enzyme, leading to a metabolic consequence that may be fatal.

inbreeding The mating of related individuals.

incomplete dominance A pair of alleles, neither of which displays dominance, the phenotype of a heterozygote being intermediate between the phenotypes of the two homozygotes.

incomputable Something that cannot be characterized by a program that always halts.

incomputable number A real number with an infinite decimal (or binary) expansion that cannot be enumerated by any universal computer.

in-degree The number of edges in a directed graph that end at (point into) the specified vertex.

indel A position in an alignment between two DNA sequences where an insertion or deletion has occurred.

independent assortment The principle that is unlinked loci, the alleles of one locus segregate independently of the alleles of the other.

index A database feature used to locate data quickly within a table.

Indexed Flat Files (IFFs) Partially structured databases, which may include a thesaurus (adding the ability to search synonyms) or other basic search tools. IFFs, meanwhile, allow users to interactively navigate among entries in several different databases by means of hypertext links. IFFs do not, however, allow true database integration, and gathering information from these types of files is often haphazard: because the data are not really structured, researchers may end up with many incorrect matches to their queries. The principal advantage of this technology is that it is cheap and easy to understand.

inducer A molecule that induces expression of a gene or operon by binding to a repressor protein and preventing the repressor from attaching to the operator.

induction The switching of cells between pathways under the influence of an adjacent group of cells. It is possible to generate several different cells through a series of inductions between a limited number of cell types.

inference The logical process by which new facts are derived from known facts by the application of inference rules.

inferred tree A rooted tree obtained by phylogenetic analysis; a phylogenetic tree based on empirical data pertaining to extant taxa.

informatics The study of the application of computer and statistical techniques to the management of information. In genome projects, informatics includes the development of methods to search databases quickly, to analyse DNA sequence information, and to predict protein structure from DNA sequence data.

information content The number of bits needed to uniquely specify a message, where one bit is able to distinguish between two equally likely things.

information superhighway A popular term associated with the Internet, used to describe its role in the global mass transportation of information.

information theory Initially developed by Claude Shannon, describes the amount of data that can be transmitted across a channel given specific encoding techniques and noise in the signal.

informed consent An individual willingly agrees to participate in an activity after first being adviced of the risks and benefits.

inheritable Able to be passed genetically from parent to offspring.

inherited Genetic material transmitted from parents through biological processes.

inhibitor-resistant mutant An organism with a mutation that is able to resist the toxic effects of an antibiotic or other type of inhibitor.

inhibitory A neural synapse or weight that is negative such that activity in the source neuron encourages inactivity in the connected neuron.

initiation The first step in translation, that occurs when a messenger RNA molecule, a ribosomal subunit, and a transfer RNA molecule carrying the first amino acid bind together to form a complex.

initiation complex The complex of proteins that initiates transcription and also this complex initiates translation.

initiation factor A protein that plays an ancillary role during initiation of translation.

initiation of transcription The assembly upstream of a gene of the complex of proteins that will subsequently copy the gene into RNA.

initiation region A region of eukaryotic chromosomal DNA within which replication initiates at positions that are not clearly defined.

initiator sequence (Inr) A component of the RNA polymerase II core promoter sequence.

initiator tRNA The tRNA, aminoacylated with methionine in eukaryotes or N-formylmethionine in bacteria, that recognizes the initiation codon during protein synthesis.

inner product For two vectors of the same dimensionality, the sum of the pairwise products of the two vector components.

inode The internal representation of a file in the UNIX operating system.

inosine A modified version of guanosine, sometimes found at the wobble position of an anticodon.

in-phase overlapping The condition in which two or more proteins are translated in the same reading frame.

insertion A chromosome abnormality in which a piece of DNA is incorporated into a gene and thereby disrupts the gene's normal function.

insertional editing A less extensive form of pan-editing that occurs during processing of some viral RNAs.

insertion mutation   A mutation that arises by insertion of one or more nucleotides into a DNA sequence.

insertion sequence (IS)   A short DNA sequence that acts as a simple transposable element found in bacteria. Insertion sequences have two major characteristics: they are small relative to other transposable elements (generally around 700 to 2500 bp in length) and only code for proteins implicated in the transposition activity (they are thus different from transposons, which also carry accessory genes such as antibiotic resistance genes). These proteins are usually transpose, which catalyses the enzymatic reaction allowing the IS to move, and one regulatory protein which either stimulates or inhibits the transposition activity. The coding region in an insertion sequence is usually flanked by inverted repeats. It is also called as an insertion sequence element, or an IS element.

in silico   Experimental process performed on a computer and not by bench research.

***in silico* biology**   The use of computers to simulate or analyse a biological experiment.

***in situ* hybridization**   A variation of the DNA/RNA hybridization procedure in which the denatured DNA is in place in the cell and is then challenged with RNA or DNA extracted from another source. Hybridization of a labelled probe to its target which has been fixed for visualization by microscopy.

integer   A whole number, i.e., a number without fractional parts.

integrase A   Type I topoisomerase that catalyses the insertion of the lambda genome into *Escherichia coli* DNA.

integrated databases   Integration [of databases] typically is accomplished by creating small, object-oriented software elements, or "wrappers" that let a single overlaying, often browser-like, desktop application interact with all the pieces. The original separate systems are intact and functional, and new ones can be added, while the underlying complexity is transparent to users. There are still many challenges, but computing environments are becoming more unified, flexible and expandable. Information in OMIM [Online Mendelian Inheritance in Man] and the published working draft of the International Human Genome Sequencing Consortium has been facilitated by ties to NCBI's RefSeq and LocusLink databases.

integration   The physical insertion of DNA into the host cell genome. The process is used by retroviruses where a specific enzyme catalyses the process or can occur at random sites with other DNA (e.g. transposons).

integration (of databases) Allows researchers to increase the value that they get from the data, because it increases the base of information that they can access and allows for more robust searching.

intein An internal segment of a polypeptide that is removed by a splicing process after translation.

intein homing The conversion of a gene coding for a protein that lacks an intein into one coding for an intein-plus protein, catalysed by the spliced component of the intein.

Intellectual Property (IP) Legal property rights over creations of the mind, both artistic and commercial, and the corresponding fields of law.   Under Intellectual Property Law, owners are granted certain exclusive rights to a variety of intangible assets, such as musical, literary, and artistic works; ideas, discoveries and inventions; and words, phrases, symbols, and designs. Common types of intellectual property include copyrights, trademarks, patents, industrial design rights and trade secrets.   The majority of intellectual property rights provide creators of original works eco-nomic incentive to develop and share ideas through a form of temporary monopoly.

intercalating agent A compound that can enter the space between adjacent base pairs of a double-stranded DNA molecule, often causing mutations.

interference The event which inhibits the chances of another crossover event.

interferons A protein released by cells in response to viral infection. It activates the synthesis and secretion of antiviral proteins.

internal environment In multicellular organisms, the aqueous environment that is outside the cells but inside the body.

internal gene duplication Repeated sequences within a gene that have been derived from duplications involving less than the entire gene sequence.

internal node A branch point within a phylogenetic tree, representing an organism or DNA sequence that is ancestral to those being studied.

Internal Ribosome Entry Site (IRES) A nucleotide sequence that enables the ribosome to assemble at an internal position in some eukaryotic mRNAs.

International Centre for Genetic Engineering and Biotechnology (ICGEB) "An international organization dedicated to advanced research and training in molecular biology and biotechnology, with special regard to the needs of the developing world," in Trieste, Italy.

International Nucleotide Data Banks The coordinated efforts of the GenBank, EMBL, and DDBJ sequence databases.

International Organization for Standardization (ISO) An international standard-setting body composed of representatives from various national standards organizations. Founded on 23 February 1947, the organization promulgates worldwide proprietary industrial and commercial standards. It is headquartered in Geneva, Switzerland. While ISO defines itself as a non-governmental organization, its ability to set either standards that often become law, through treaties or national standards, makes it more powerful than most non-governmental organizations. In practice, ISO acts as a consortium with strong links to governments.

International Society for Computational Biology (ISCB) A society "dedicated to advancing the scientific understanding of living systems through computation; our emphasis is on the role of computing and informatics in advancing molecular biology."

internet A network of linked computer networks used to transmit files and messages.

interphase The period in the cell cycle between cell division when DNA is replicated in the nucleus. This phase is followed by mitosis.

InterPro A database of protein families, domains and functional sites in which identifiable features found in known proteins can be applied to unknown protein sequences.

Interspersed Repeat Element PCR (IRE-PCR) *See* repititive DNA PCR.

interspersed repetitive DNA DNA sequences that are present many thousands of times in each haploid genome, which are widely scattered throughout the genome and which are not particularly concentrated at any point.

intracellular domain A domain family that is most prevalent in proteins within the cytoplasm.

intracellular signalling The communication of a molecular message from the surface of the cell to the nucleus via the participation of a series of molecules, including receptors, enzymes, proteins, and small molecules. The end result of the signalling process is the up- or down-regulation of a particular series of genes that may be involved in cell growth, division or differentiation.

intranet A computer network, based upon World Wide Web and Internet technologies, but whose scope is limited to an organization.

intrinsic terminator A position in bacterial DNA where termination of transcription occurs without the involvement of Rho.

intron The portion of a DNA sequence which interrupts the protein-coding sequences of the gene. Most introns begin with the nucleotides GT and end with the nucleotides AG.

introns early The hypothesis that introns evolved relatively early and are gradually being lost from eukaryotic genomes.

introns late The hypothesis that introns evolved relatively late and are gradually accumulating in eukaryotic genomes.

invariant repitition The existence of repeated DNA segments that are identical or nearly identical in sequence to one another.

inversion A structural aberration in a chromosome in which the order of several genes is reversed from the normal order. A pericentric inversion includes the centromere within the inverted region, and a paracentric inversion does not include the centromere.

inverted repeat Two identical nucleotide sequences repeated in opposite orientations in a DNA molecule.

invertible A function is invertible (with a unique inverse) if the output uniquely determines the input (i.e., it is one-to-one) and the set of legal outputs is equal to the set of legal inputs.

in vitro Experimental process performed in a tube or Petri dish and not in a living cell or organism. Literally translated as "in glass".

***in vitro* mutagenesis** Techniques used to produce a specified mutation at a predetermined position in a DNA molecule.

in vivo Experimental process performed in live cells or organisms, compared to *in vitro* or *in silico*.

IP address The unique, numeric address of a computer host on the Internet.

Irish National Centre for BioInformatics (INCBI) The Irish national node of EMBnet. INCBI is located in the Genetics Department of Trinity College, Dublin.

irrational number A real number that cannot be represented as a fraction.

isoaccepting tRNAs Two or more tRNAs that are charged with the same amino acid.

isochore A large region of DNA (greater that 3 kb) with a high degree uniformity in G–C and C–G (collectively GC) which tends to have more genes, higher local melting or denaturation temperatures, and different flexibility, i.e., isochores are largely homogeneous in GC content in contrast to the heterogenity of the entire genome.

isochromosome A chromosome in which one arm has been lost and has been replaced by a duplicate of the other arm.

isoelectric focusing A technique for separating proteins on the basis of their charge.

isoelectric point The pH at which a molecule has no net electrical charge and therefore does not move in an electric field.

isoenzyme An enzyme performing the same function as another enzyme but having a different set of amino acids.

isoform One of several forms of the same protein that differs in its amino acid sequence. Isoform is produced by different genes or alternative splicing of mRNA.

isogene The variants or mutations within the genes of a patient population.

isoleucine An α-amino acid with the chemical formula $HO_2CCH(NH_2)$ $CH(CH_3)$ CH_2CH_3. It is an essential amino acid, which means that humans cannot synthesize it, so it must be part of our diet. Its codons are AUU, AUC and AUA.

isoprene Small unsaturated hydrocarbon containing five carbon atoms.

isoschizomers Two different restriction enzymes which recognize and cut DNA at the same recognition site, e.g. *Sma*I and *Xma*I both recognize and cut the sequence CCCGGG.

isotope One of two or more atoms that have the same atomic number but different neutrons.

isotropic temperature factor A consequence of the dynamic disorder in the crystal caused by the temperature-dependent vibration of the atoms in the structure.

isotype An antigenic determinant that distinguishes the constant regions of the different heavy-chain classes and light-chain types.

isozyme Two or more enzymes capable of catalysing the same reaction but varying in their specificity due to differences in their structures and hence their efficiencies under different environmental conditions.

Iterated Functional System (IFS) A method for constructing a fractal by iterating a vector quantity through an affine equation that is randomly selected on each iteration.

iteration A series of steps in an algorithm whereby the processing of data is performed repetitively until the result exceeds a particular threshold. Iteration is often used in multiple sequence alignments whereby each set of pairwise alignments are compared with each other, starting with the most similar pairs and progressing to the least similar, until there are no longer any sequence-pairs remaining to be aligned.

iterative search A search procedure that is repeated, usually with increasing sensitivity in each round.

J

Janus Kinase (JAK) A type of kinase that plays an intermediary role in some types of signal transduction involving STATs.

Jarvis-Patrick cluster algorithm A non-hierarchical clustering method for conformational analysis of molecules that uses a "nearest neighbour" approach to look for clusters of conformations that are the shortest distance away from each other.

Java An object-oriented programming language developed by Sun Microsystems to allow programs to be run on any computer.

jelly-roll A variant of Greek key topology with both ends of a sandwich or a barrel fold being crossed by two interstrand connections.

join An SQL statement used to combine the data contained in two relational database tables based upon a common attribute.

joule Standard unit of energy in the metre-kilogram system.

Julia set A set of complex numbers that do not diverge if iterated an infinite number of times via a simple equation.

***jun* oncogene** An oncogene that produces the Jun protein, a transcriptional regulator (with a leucine zipper motif) that recognizes AP-1 sites in DNA.

junctional diversification The random loss and gain of nucleotides at immunoglobin V(D)J joining sites.

junk DNA The excess DNA that is present in the genome beyond that required to encode proteins. A misleading term since these regions are likely to be involved in gene regulation, and other as yet unidentified functions. The 97% of the human genome that is repetitive sequence and does not contain exons.

K

K A statistical parameter used in calculating BLAST scores that can be thought of as a natural scale for search space size. The value K is used to convert a raw score (S) into a bit score (S´).

Kappa particle An intracellular parasite in *Paramecium* that releases a substance capable of killing sensitive cells.

karyogram The entire chromosome complement of a cell, with each chromosome described in terms of its appearance at metaphase.

karyotype The chromosomal constitution of an individual.

KEGG Kyoto Encyclopedia of Genes and Genomes. An effort to computerize the current knowledge of molecular and cellular biology in terms of the information pathways that consist of interacting molecules or genes and to provide links from the gene catalogs produced by genome sequencing projects.

Keratins A family of fibrous structural proteins; tough and insoluble, they form the hard but non-mineralized structures found in reptiles, birds, amphibians, and mammals. As biological materials they are rivalled in toughness only by chitin.

kernel The lowest level of the UNIX operating system.

key An attribute used to sort and/or identify data in some manner.

keyword A word which is identified as being particularly informative about an object. In a sequence, a keyword often relates to the identity of a gene or the function of the gene product. References often have a list of keywords that are Medline MeSH terms. Keywords are good to use in text searches.

K-homology domain A type of RNA-binding domain.

kilobase (kb) A unit of length for DNA fragments that equals 1000 nucleotides.

kilocalorie A unit of heat energy equals to 1000 calories.

kinase A type of enzyme that transfers phosphate groups from high-energy donor molecules, such as ATP, to specific substrates. The process is referred to as phosphorylation. An enzyme that removes phosphate groups is known as a phosphatase.

kindred A very extended pedigree, i.e., a group of relatives.

kinetochore The part of the centromere to which spindle microtubules attach.

kissing complex The bipartite structure formed by base pairing between a small regulatory RNA and a messenger RNA.

Klinefelter syndrome The clinical features of human males with the karyotype 47, XXY.

knockout mouse Mice which have been engineered to lack a chosen gene. The gene is inactivated so-called, embryonic stem cells using the technique of homologous recombination. These cells are then introduced into an early stage embryo (blastocyst) and then it is transplanted into a recipient mouse. The subse-quent progeny lack the targeted gene in some cells. This technique is used to determine the function of the chosen gene.

knockout mutation Any mutation that completely eliminates the function of the gene. It is also termed as null mutation, loss-of-function mutation, and amorphic mutation.

Knowledge Acquisition (KA) A sub-speciality of artificial intelligence, usually associated with developing methods for capturing human knowledge and of converting it into a form that can be used by computer.

Koch curve A fractal curve that looks like the edge of a snowflake.

Kosambi's mapping function A mapping function based on the assumption that the coincidence between crossovers is proportional to the distance between them.

Kozak consensus The nucleotide sequence surrounding the initiation codon of an eukaryotic mRNA.

Krebs cycle The central metabolic pathway found in all aerobic organisms.

kruskal's algorithm An algorithm for computing a minimum spanning tree.

ksh Korn shell command interpreter in UNIX.

ktup The word size used to make the hash table in the FASTA program.

Kuru A neurological disease caused by a prion protein.

Kyoto An external link in the Locus or Clone page to the Kyoto Encyclopedia of Genes and Genomes.

L

label A chemical tag attached to or included within a molecule to enable its subsequent identification.

lab on a chip Microdevices that allow rapid, microanalytical analysis of DNA or protein in a single, fully integrated system. Typically, these devices are miniature surfaces, made of silicon, glass or plastic, which carry the necessary microdevices (pumps, valves, microfluidic controllers, and detectors) that allow sample separation and analysis. These devices are used in drug discovery, genetic testing and separation science.

lactose operon The cluster of three genes that code for enzymes involved in utilization of lactose by *E. coli*.

lactose repressor A regulatory protein that controls transcription of the *lac* operon in response to the presence or absence of lactose in the environment.

lagging strand The strand of the double helix that is copied in a discontinuous fashion during DNA replication.

Lamarckism The view, articulated by Jean-Baptiste Lamarck, that features acquired during an organism's lifetime would be passed on to succeeding generations, leading to inheritable change in species over time.

lambda A statistical parameter used in calculating BLAST scores that can be thought of as a natural scale for scoring system. The value lambda is used in converting a raw score (S) to a bit score (S′).

lambda calculus A model of computation that is capable of universal computation. The Lisp programming language was inspired by lambda calculus.

lariat The lariat-shaped intron RNA that results from splicing a GU–AG intron.

law of the minimum States that the population growth is limited by the resource in shortest supply.

lead compound A candidate compound identified as the best "hit" (tight binder) after screening of a combinatorial (or other) compound library, that is then taken into further rounds of screening to determine its suitability as a drug.

leader segment The untranslated region of an mRNA upstream of the initiation codon.

leading strand The strand of the double helix that is copied in a continuous fashion during DNA replication.

lead optimization The process of converting a putative lead compound ("hit") into a therapeutic drug with maximal activity and minimal side effects, typically using a combination of computer-based drug design, medicinal chemistry and pharmacology.

leaky mutation A mutation that results in partial loss of a characteristic.

learning A process of adaptation by which synapses, weights of neural networks, classifier strengths, or some other set of adjustable parameters are automatically modified so that some objective is more readily achieved.

length abridgement The gradual shortening of pseudogenes during evolution that is caused by an excess of deletions over insertions.

lethal mutation A mutation that results in death of the cell or organism.

leucine (Leu, L) An α-amino acid with the chemical formula $HO_2CCH(NH_2)$ $CH_2CH(CH_3)_2$. It is an essential amino acid, which means that humans cannot synthesize it. Its codons are UUA, UUG, CUU, CUC, CUA, and CUG.

leucine zipper Protein motif which binds to a DNA in which 4–5 leucines are found at 7 amino acid intervals. This motif is present typically in transcription factors and other proteins that bind DNA.

lexicon A set of terms used to describe a particular activity or field of study. In bioinformatics, a lexicon refers to a pre-defined list of terms that together completely define the contents of a particular database.

library A collection of expressed genes from a specific tissue sample, and their annotations, i.e., a large collection of compounds, peptides, cDNAs or genes which may be screened in order to isolate cognate molecules.

life A set of rules for a cellular automaton that operates on a two-dimensional grid.

ligand Any small molecule that binds to a protein or receptor and also it is a cognate partner of many cellular proteins, enzymes and receptors.

ligand binding When a chemical messenger is released by a cell it must bind to the receptor on its target cell to receive the necessary chemical information.

ligation Formation of a phosphodiester bond to link two adjacent bases separated by a nick in double-stranded DNA.

limit cycle A periodic cycle in a dynamical system such that previous states are returned to repeatedly.

linear Having only a multiplicative factor; a function that can be graphed as a line.

linearly separable Two classes of points are linearly separable if a linear function exists such that one class of points resides on one side of the hyperplane (defined by the linear function), and all points in the other class are on the other side.

linkage The association of genes (or genetic loci) on the same chromosome, i.e., genes that are linked together tend to be transmitted together.

linkage analysis The process used to study the genotype variations between affected and healthy individuals wherein specific regions of the genome that may be inherited with, or "linked" to disease are determined.

linkage disequilibrium The occurrence, on the same chromosome, of some combinations of alleles of closely linked genes more often than would be predicted by chance.

linkage map A map of the relative positions of genetic loci on a chromosome, determined on the basis of how often the loci are inherited together.

linker DNA The DNA that links nucleosomes. It is the "string" in the beads-on-a-string model for chromatin structure.

linker histone A histone, such as H1, that is located outside of the nucleosome core octamer.

lint A UNIX program that can detect bugs, portability problems, and other possible errors in C programs.

Linux A free, UNIX-type operating system originally created by Linus Torvalds.

Lisp (List Processor) A programming language designed to manipulate lists, inspired by lambda calculus.

lithosphere The solid outer layer of the earth. It includes both the land area and the land beneath the oceans and other water bodies.

little-o notation A theoretical measure of the execution of an algorithm, usually the time or memory needed, given the problem size n, which is usually the length of the input string.

Lm distance A family of functions for distance measurement.

local alignment An alignment of some portion of two nucleic acids or proteins sequences.

Local Area Network (LAN) A network that connects computers in a small, defined area, such as the offices in a single wing or a group of buildings.

localize Determination of the original position locus of a gene or other marker on a chromosome.

localized repeated sequence Tandemly arrayed, repetitive sequences, usually made up of short-simple repeated motifs (e.g. satellite DNA).

local maximum The top of a peak and a point in a search space such that all nearby points are lower.

local minimum The bottom of a valley and a point in a search space such that all nearby points are higher.

local similarity An alignment technique where the subunits of two sequences that exhibit most similarity are aligned. In biology, this is normally termed as local alignment.

lock A mechanism used to assure that concurrent database transactions are handled appropriately.

locus The specific position occupied by a gene on a chromosome. At a given locus, any one of the variant forms of a gene may be present. The variants are said to be alleles of that gene.

Locus Control Region (LCR) A DNA sequence that maintains a functional domain in an open, active configuration.

logarithmic A scale of measurement that uses the logarithm of a physical quantity instead of the quantity itself.

logistic growth model A model of population growth in which the population initially grows at an exponential rate until it is limited by some factor; then, the population enters a slower growth phase and eventually stabilizes.

logistic map The simplest chaotic system that works in discrete time and is defined by the map $x(t) = 4r\, x(t)(1-x(t))$.

log-odds A scoring system in which the values are the logarithm of the relative probability (odds) of a comparison being due to homology or being due to chance.

log score A statistical measure of linkage determined by pedigree analysis.

Long Branch Attraction (LBA) A phenomenon in phylogenetic analyses (most commonly those employing maximum parsimony) when rapidly evolving lineages are inferred to be closely related, regardless of their true evolutionary relationships.

Long Interspersed Nuclear Element (LINE) A type of genome-wide repeat, often with transposable activity. LINE-1 is a type of human LINE.

long patch repair A nucleotide excision repair process of *Escherichia coli* that results in excision and resynthesis of up to 2 kb of DNA.

long-range restriction mapping Restriction enzymes are proteins that cut DNA at precise locations. Restriction maps depict the positions on chromosomes of restriction enzyme cutting sites.

loop In a graph, an edge that connects a vertex to itself.

Lorenz system A system of three differential equations that was the first concrete example of chaos and a strange attractor.

loss-of-function mutation A mutation that reduces or terminates the activity of a protein.

Lotka–Volterra system A two-species predator-prey system that in its simplest form can display only fixed points or limit cycles. More complicated versions with three or more species can yield chaos.

Low-Complexity Region (LCR) A region of biased composition including homopolymeric runs, short-period repeats, and more subtle over representation of one or a few residues. The SEG program is used to mask or filter the LCRs in amino acid queries. The DUST program is used to mask or filter the LCRs in nucleic acid queries.

lowly repetitive gene Genes appearing in only a few copies in the haploid genome.

ls UNIX command for listing the contents of a directory.

L-system A method of constructing a fractal and it is also a model for plant growth.

LTR element A type of genome-wide repeat typified by the presence of long terminal repeats (LTRs).

Lyon hypothesis Idea proposed by Mary Lyon that mammalian females inactivate one of the two X chromosomes during early embryogenesis.

lysine (Lys, K) An α-amino acid with the chemical formula $HO_2CCH(CH_2)_4NH_2$. This amino acid, is an essential amino acid which means that humans cannot synthesize it. Its codons are AAA and AAG.

lysogenic pathway The type of bacteriophage infection that involves integration of the phage genome into the host genome.

lytic pathway The type of bacteriophage infection that involves lysis of the host cell immediately after the initial infection, with no integration of the phage DNA molecule into the host genome.

M

M13 A single-stranded DNA bacteriophage.

machine language The set of commands that a computer's processor uses to carry out functions.

Machine Learning (ML) Sub-speciality of artificial intelligence concerned with developing methods for software to learn from experience, or to extract knowledge from examples in a database.

Mackey–Glass System A delay differential equation that can display a wide variety of behaviours via an adjustable delay term, tau.

macrochromosome One of the largest, gene-deficient chromosomes seen in the nuclei of chickens and various other species.

macroevolution The combination of events associated with the origin, diversification, extinction, and interactions of organisms which produced the species that currently inhabit the earth.

macrorestriction map Map depicting the order of and distance between sites at which restriction enzymes cleave chromosomes.

MADS box A DNA-binding domain found in several transcription factors involved in plant development.

maintenance methylation Addition of methyl groups to positions on newly synthesized DNA strands that correspond with the positions of methylation on the parent strand.

major groove The larger of the two grooves that spiral around the surface of the B-form of DNA.

Major Histocompatibility Complex (MHC) A mammalian multigene family coding for cell surface proteins and including several multiallelic genes.

Mandelbrot set An extremely complex fractal that is related to Julia sets in the way that it is constructed and by the fact that it acts as a sort of index to the Julia sets.

Manhattan distance The distance between two points measured along axes at right angles.

map 1. A chart showing the positions of genetic and/or physical markers in a genome. 2. Program developed by Huang which computes a multiple global alignment using an iterative pairwise method.

MAP kinase A signal transduction pathway.

mapping The process of determining the positions of genes and the distances between them on a chromosome. This is accomplished by indentifying unique genome markers (ESTs, STSs, etc.) and localizing these to specific sites on the chromosome. There are three types of DNA maps: physical maps, genetic maps, and cytogenetic maps. The types of markers identified differentiate the map produced.

mapping population The group of related organisms used in constructing a genetic map.

mapping reagent A collection of DNA fragments spanning a chromosome or the entire genome and used in STS mapping.

map position The point at which a marker is found on a genetic map, or at which a clone is found on a cytogenetic map.

map unit A measure of genetic distance between two linked genes that corresponds to a recombination frequency of 1%. 1 map unit = 1 centi-morgan.

marker A physical location on a chromosome which can be reliably monitored during replication and inheritance. Markers on the human transcript map are all STSs.

Markov chain A finite state machine with probabilities for each transition, that is, a probability that the next state is sj given that the current state is si, i.e., any multivariate probability density whose independence diagram is a chain.The variables are ordered, and each variable "depends" only on its neighbours in the sense of being conditionally independent of the others. Markov chains are an integral component of Hidden Markov models.

Markov model A statistical model for sequences in which the probability of each letter depends on the letters that precede it.

masking The removal of repeated or low-complexity regions from a sequence in order to improve the sensitivity of sequence similarity searches performed with that sequence. Also known as filtering.

mass extinction A time during which extinction rates are generally accelerated so that more than 50% of all species then living become extinct; results in a marked decrease in the diversity of organisms.

mass spectrometry An instrument used to identify chemicals in a substance by their mass and charge.

match In sequence alignment, the existence of the same base in a homologous position in both sequences.

maternal-effect gene A *Drosophila* gene that is expressed in the parent and whose mRNA is subsequently injected into the egg, after which it influences development of the embryo.

mating type The equivalent of male and female for a eukaryotic microorganism.

Mating-type interconversion Phenomenon in homothallic yeast in which cells switch mating type as a result of the transposition of genetic information from an unexpressed cassette into the active mating-type locus.

mating-type switching The ability of yeast cells to change mating type by gene conversion.

matrix A rectangular two-dimensional array of numbers that can be thought of as a linear operator on vectors.

Matrix-Associated Region (MAR) An AT-rich segment of a eukaryotic genome that acts as an attachment point to the nuclear matrix.

matter Anything that has mass and occupies space.

matter cycling The flow of matter through various organisms and the physical environment of an ecosystem.

maturase An enzyme, coded by a gene in an intron, thought to be involved in splicing.

maturation The formation of mRNA from pre-mRNA.

maximum parsimony method A method for construction of phylogenetic trees requiring the fewest substitutions from among all possible phylogenetic trees.

mean The arithmetical average of a collection of numbers and also the centre of a Gaussian distribution.

meander A simple topology of a beta-sheet where any two consecutive strands are adjacent and antiparallel.

mediator A protein complex that forms a contact between various transcription factors and the C-terminal domain of the largest subunit of the yeast RNA polymerase II.

medical bioinformatics Linking clinical data to patient gene profiling. This data covers haplotyping, genotyping, population genomics, gene expression profiling, particularly for use in diagnosis, prognosis and therapeutic stratification of patients.

Medical College of Wisconsin (MCW) A national, private, academic institution dedicated to leadership and excellence in biomedical education, research, patient care and service.

Medical Subject Headings (MeSH) A controlled vocabulary used to index the medical literature, developed by the U.S. National Library of Medicine, and applied in applications such as MedLine and PubMed.

MedLine The bibliographic database of more than 12 million citations in the fields of medicine, nursing, dentistry, veterinary medicine, the health care system, and the preclinical sciences, compiled by the U.S. National Library of Medicine (NLM).

megabase (Mb) Unit of length for DNA fragments equal to 1 million nucleotides and roughly equal to 1 cM.

megabyte (Mb) 1 million bytes.

meiosis A process within the cell nucleus that results in the reduction of the chromosome number from diploid (two copies of each chromosome) to haploid (a single copy) through two reductive divisions in germ cells.

melting Denaturation of a double-stranded DNA molecule due to breakdown of the hydrogen bonds between paired nucleotides.

melting temperature (T_m) The temperature at which the two strands of a double-stranded nucleic acid or base-paired hybrid detach due to dissolution of the hydrogen bonds that hold the pairs together.

meme A unit of cultural information that represents a basic idea that can be transferred from one individual to another, and subjected to mutation, crossover and adaptation.

Mendelian inheritance One method in which genetic traits are passed from parents to offspring.

Mendelian segregation (Mendel's first law). The principle states that the two different alleles of a gene pair in a heterozygote separate from each other during meiosis to produce two kinds of gametes in equal ratios, each bearing a different allele.

message The basic unit of information in a classifier system that is stored in the message list.

message list The portion of a classifier system that retains information in the form of messages.

messenger RNA (mRNA) A single-stranded molecule of ribonucleic acid that directs protein production by serving as a template for protein synthesis, i.e., the complementary RNA copy of DNA formed from a single-stranded DNA template during transcription that migrates from the nucleus to the cytoplasm where it is processed into a sequence carrying the information to code for a polypeptide domain. The first codon in an mRNA sequence is almost always AUG.

meta-analysis Pooled statistical analysis of results from several individual statistical analyses of different experiments, searching for statistical significance which is not possible within the smaller sample sizes of individual studies.

metabolic pathway The biochemical reaction networks that make up the metabolism of an organism.

metabolism The sum of all chemical reactions (energy exchanges) in cells.

metabolomics An area of functional genomics, which studies changes in the expression of small organic molecules in biological systems.

metacentric A chromosome whose centromere lies in the middle.

metadata Information about the structure of the data in a database such as the origin, size, formatting or other characteristics of data items.

metaphase A stage in mitosis or meiosis during which the chromosomes are aligned along the equatorial plane of the cell.

metaphase chromosome A chromosome at the metaphase stage of cell division, when the chromatin takes on its most condensed structure and features such as the banding pattern can be visualized.

Methionine (Met, M) An amino acid coded for by the initiation codon; all polypeptides begin with methionine, although post-translational reactions may remove it.

methylation The addition of –CH$_3$ (methyl) groups to a target site. Typically such addition occurs in the cytosine bases of DNA.

Methyl-CpG-Binding Protein (MeCP) A protein that binds to methylated CpG islands and may influence acetylation of nearby histones.

metric space A set of objects on which a distance function is defined.

MGMT D-Methylguanine DNA Methyltransferase. An enzyme involved in the direct repair of alkylation mutations.

MIAME/Tox A proposal by the EBI, NIEHS-NCT and ILSI-HESI for defining the Mininum Information about Microarray Experiment requirements for Toxicogenomics. A standard for describing a microarray experiment.

microarray An array of oligo-nucleotides printed on a solid surface to identify fluorescently labelled DNA sequences.

MicroArray and Gene Expression (MAGE) A standard microarray expression data to facilitate the exchange information.

Microarray Gene Expression Database (MGED) "The Microarray Gene Expression Data (MGED) Society is an international organization of biologists, computer scientists, and data analysts that aims to facilitate the sharing of microarray data generated by functional genomics and proteomics experiments."

microbial genetics The study of genes and gene function in bacteria, archaea, and other microorganisms.

microevolution A small-scale evolutionary event such as the formation of a species from a pre-existing one or the divergence of reproductively isolated populations into new species.

microfluidics The miniaturization of chemical reactions or pharmacological assays into microscopic tubes or vessels in order to greatly increase their throughput, by placing many of them side-by-side in an array.

microinjection A technique for introducing a solution of DNA into a cell using a fine microcapillary pipette.

microsatellite A specific sequence of DNA bases or nucleotides which contains mono-, di-, tri-, or tetra tandem repeats. For example,

GGGGGGGG is a $(G)_8$

ACACAC is referred to as $(AC)_3$

ATCATCATCATCATC would be referred to as $(ATC)_5$

ATCTATCT would be referred to as $(ATCT)_2$

Microsatellites are also called simple sequence repeats (SSR), short tandem repeats (STR), or variable number tandem repeats (VNTR).

migration In population genetics, the movement of individuals or genes among populations.

mimetics Compounds that mimic the function of other molecules via their high degree of structural (conformational) similarity, and hence physiochemical properties.

MIM number The catalog number for a gene or Mendelian character, as listed in Victor McKusick's Mendelian Inheritance in Man.

minichromosome One of the smallest, gene-rich chromosomes seen in the nucleus of chickens and various other species.

minigene A pair of exons carried by a cloning vector used in the exon-trapping procedure.

minimal medium A medium that provides only the minimum nutritional requirements for growth of a microorganism.

Minimum Spanning Tree (MST) A tree in a weighted graph that contains all of the graph's vertices, where the sum of the weights along the edges is minimal.

minisatellite A type of simple sequence length polymorphism comprising tandem copies of repeats that are a few tens of nucleotides in length.

minor groove The smaller of the two grooves that spiral around the surface of B-DNA.

mismatch A position in a double-stranded DNA molecule where base-pairing does not occur because the nucleotides are not complementary.

mismatch repair A DNA repair process that corrects mismatched nucleotide pairs by replacing the incorrect nucleotide in the daughter polynucleotide.

missense mutation An alteration in a nucleotide sequence that converts a codon for one amino acid into a codon for a second amino acid.

mitochondrial DNA The DNA of the mitochondrial genome.

mitochondrial genome The genome present in the mitochondria of an eukaryotic cell.

mitochondrial inheritance Inheritance of a character encoded in the mitochondrial genome.

mitochondrion   A self-replicating, membrane-bound cytoplasmic organelle of eukaryotic cells that metabolize glucose to produce energy-bearing NADH and ATP.

mitosis   The process of nuclear division in cells that produces daughter cells that are genetically identical to each other and to the parent cell.

mixed strategy   A strategy that uses randomness to select different actions in identical circumstances with different probabilities.

model   Any representation of a real object or phenomenon, or template for the creation of an object or phenomenon.

model-based reasoning An approach to the development of expert systems that uses formally defined models of systems, in contrast to more superficial rules or heuristics.

modelling A process whereby the three-dimensional architecture of biological molecules is interpreted (or predicted), visually represented, and manipulated in order to determine their molecular properties. The use of statistical analysis, computer analysis, or model organisms to predict outcomes of research.

model of computation   An idealized version of a computing device that usually has some simplifications such as infinite memory.

model organism   An organism that is relatively easy to study, and can be used to obtain information that is relevant to the biology of a second organism that is more difficult to study.

moderately repetitive genes Genes with a moderate number of copies in the haploid genome.

modification assay   A range of techniques used for locating bound proteins on DNA molecules.

modification interference A technique used to identify nucleotides involved in the interactions with a DNA-binding protein.

module   In globular proteins, a   structurally   independent, stable and compact spatial unit that can be distinguished from all other parts, usually consisting of a continuous stretch of amino acids.

modus ponens A rule of logical inference that if proposition A is true, and A implies B, then B is true.

molecular bioinformatics Conceptualizing biology in terms of molecules (in the sense of physical- chemistry) and then applying "informatics" techniques (derived from disciplines) such as applied maths, CS [computer science] and statistics to understand and organize the information associated with these molecules on a large-scale.

molecular biology The study of the structure, function and make-up of biologically important molecules.

molecular chaperone A protein that helps other proteins to fold.

molecular clock The hypothesis that nucleotide or amino acid substitutions occur at more or less fixed rates over evolutionary time, like the slow ticking of a clock.

molecular combing A technique for preparing restricted DNA molecules for optical mapping.

molecular evolution The gradual changes that occur in genomes over time due to the accumulation of mutations and structural rearrangements resulting from recombination and transposition.

molecular farming The development of transgenic animals to produce human proteins for medical use.

molecular genetics The study of macromolecules important in biological inheritance.

molecular life sciences An area of research comprising molecular biology, biochemistry and cell biology, as well as some aspects of genetics and physiology.

molecular medicine The treatment of injury or disease at the molecular level.

molecular modelling The prediction of a molecule's three-dimensional shape can be estimated from sequence data of previously identified molecular shapes.

Molecular Modelling Database (MMDB) A database of macromolecular 3D structures, as well as tools for their visualization and comparative analysis. It contains experimentally determined biopolymer structures obtained from the Protein Data Bank (PDB).

molecular phylogenetics A set of techniques that enable the evolutionary relationships between DNA sequences to be inferred by making comparisons between those sequences.

monogenic disorder A disorder caused by mutation of a single gene.

monohybrid cross A sexual cross in which the inheritance of one pair of alleles is followed.

monomer A single unit of any biological molecule or macromolecule, such as an amino acid, nucleic acid, polypeptide domain, or protein.

monomorphic A population in which virtually all individuals have the same allele at a locus.

monophyletic Sharing a common ancestor.

monosomy Having only one copy of one of the chromosomes.

monotonic The property of a function that is always strictly increasing or strictly decreasing, but never both.

monovalent Having one binding site; strictly, an atom with only one free electron available for binding in its highest energy shell.

monozygotic twins Two individuals derived from a single fertilized egg and therefore genetically identical.

Monte Carlo simulation A method of calculating significance where the analysis in question is repeated using randomized or permuted sequences, in order to determine the expected scores for unrelated (random) sequences.

morbid map A diagram showing the chromosomal location of genes associated with disease.

mortality The average probability of an individual of a given genotype to die before reaching a certain age.

mosaic An individual composed of more than one genetically distinguishable cell population derived from a single zygote.

mosaic evolution A pattern of evolution where all features of an organism do not evolve at the same rate. Some characteristics are retained from the ancestral condition while others are more recently evolved.

motif A pattern of DNA sequence that is similar for genes of similar function. Also a pattern for protein primary structure (sequence motifs) and tertiary structure that is the same across proteins of similar families.

MPEG A family of standards used for coding audio-visual information (e.g. movies, video, music) in a digital compressed format.

M phase A stage of the cell cycle when mitosis or meiosis occurs.

M13 template A piece of DNA that has been cloned into the M13 cloning vector and can be used for sequencing.

multicopy A gene, cloning vector or other genetic element that is present in multiple copies in a single cell.

multicysteine zinc finger A type of zinc finger DNA-binding domain.

multifactorial inheritance The occurrence of a phenotype as a result of the action of more than one gene and/or of environmental factors.

multifurcation A graphical representation of an unknown branching order in a phylogenetic tree involving three or more taxa.

multigene family A set of genes derived by duplication of an ancestral gene, followed by independent mutational events resulting in a series of independent genes either clustered together on a chromosome or dispersed throughout the genome.

multigraph A graph where one or more pairs of vertices are connected by multiple edges.

Multilayer Perceptron (MLP) A type of feed-forward neural network that is an extension of the perceptron in that it has at least one hidden layer of neurons.

multimedia Computer systems or applications that are able to manipulate the data in multiple forms, including still and video images, sound and text.

multiple alignment An alignment of three or more nucleotide sequences, with gaps (spaces) inserted in the sequences such that residues with common structural positions and/or ancestral residues are aligned in the same column of the multiple alignment. ClustalW is one of the most widely used multiple sequence alignment programs.

multiple alleles The different alternative forms of a gene that has more than two alleles.

Multiple Reduction Copy Machine (MRCM) An algorithm that can be used to make affine fractals.

multiple sequence alignment An alignment of three

or more sequences with gaps inserted in the sequences such that residues with common structural positions and/or ancestral residues are aligned in the same column.

multiple substitution The successive occurrence of two or more substitutions at the same nucleotide site in a DNA sequence.

multiplexing A sequencing approach that uses several pooled samples simultaneously, greatly increasing sequencing speed.

multiplex sequencing Approach to high-throughput sequencing that uses several pooled DNA samples run through gels simultaneously and then separated and analysed.

multipoint cross A genetic cross in which the inheritance of three or more markers is followed.

multiregional evolution A hypothesis that holds the modern humans in the Old World are descended from *Homo erectus* populations that left Africa more than one million years ago.

murine An organism in the genus *Mus*; a rat or mouse.

mutagen A chemical or physical agent that can cause a mutation in a DNA molecule.

mutagenesis Treatment of a group of cells or organisms with a mutagen as a means of inducing mutations.

mutagenicity The capacity of a chemical or physical agent to cause permanent genetic alterations.

mutant A cell or organism that possesses a mutation; a new variant form of a gene.

mutant phenotype A phenotype which is different from the wild type phenotype and which is caused by the possession of one or more mutant alleles.

mutasome A protein complex that is constructed during the SOS response of *Escherichia coli*.

mutation A spontaneous or induced change in the nucleotide sequence of a DNA molecule.

Mutation Data Matrix (MDM) The most common scoring system for proteins; log-odds form of the PAM-250 mutation probability matrix.

mutational bias A pattern of mutation in which the four nucleotides show different propensities to mutate or in which mutations result in a certain nucleotide more often than others.

mutation rate The number of mutations arising in an individual per nucleotide site or per gene per unit time.

mutation scanning A set of techniques for detection of mutations in DNA molecules.

mutation screening A set of techniques for determining if a DNA molecule contains a specific mutation.

mutualism A form of symbiosis in which both species benefit.

MySQL A relational database management system that implements many industry standards, including SQL and ODBC.

N

N An unknown nucleotide in DNA or RNA.

naked DNA Pure, isolated DNA devoid of any proteins that may bind to it.

NAND Negated conjunction; the opposite of the "*AND*" function.

nash equilibrium In game theory, a pair of strategies for a game such that neither player can improve his outcome by changing his strategy.

National Center for Biotechnology Information (NCBI) A unit of the National Library of Medicine (NLM), National Institutes of Health (NIH). It was established in 1988 as a national resource for molecular biology information. NCBI creates public databases, conducts research in computational biology, develops software tools for analysing genome data, and disseminates biomedical information—all for the better understanding of molecular processes affecting human health and disease NCBI is the part of NIH.

National Institute of General Medical Sciences (NIGMS) One of the US National Institutes of Health (NIH), NIGMS primarily supports basic biomedical research that is not targeted to specific diseases or disorders; NIGMS includes the Center for Bioinformatics and Computational Biology.

National Institutes of Health (NIH) An agency of the U.S. Department of Health and Human Services, Public Health Service. It supports research and training to improve health of people.

National Library of Medicine (NLM) Part of the National Institutes of Health (NIH) and includes NCBI.

National Science Foundation (NSF) An independent agency of the U.S. government that supports scientific research. Areas of interest include plant genomics, computational biology, and biocomplexity.

natural number A counting number; a positive integer.

natural selection The natural filtering process by which individuals with higher fitness are more likely to reproduce than individuals with lower fitness.

N-degron An N-terminal amino acid sequence that influences the degradation of a protein in which it is found.

negative feedback The stopping of the synthesis of an enzyme by the accumulation of the products of the enzyme-mediated reaction.

negative feedback loop A biochemical pathway where the products of the reaction inhibit the production of an enzyme that controlled their formation.

neighbor-joining method A method for construction of phylogenetic trees.

neo-Darwinism Theory of evolution that represents a synthesis of Charles Darwin's theory in terms of natural selection and modern population genetics. Neo-Darwinism asserts that evolution takes place because the environment is slowly changing, exerting a selection pressure on the individuals within a population.

nested PCR The second round amplification of an already PCR-amplified sequence using a new pair of primers which are internal to the original primers. Typically done when a single PCR reaction generates insufficient amount of product.

net input The weighted sum of incoming signals into a neuron plus a neuron's threshold value.

network A set of connected elements. For computers, any collection of computers connected together so that they are able to communicate, permitting the sharing of data or programs.

neural net An interconnected assembly of simple processing elements, units or nodes, whose functionality is based on the animal brain. The processing ability of the network is stored in the inter-unit connection strengths, or weights, obtained by a process of adaptation to, or learning from a set of training patterns. Neural nets are used in bioinformatics to map data and make predictions, such as taking a multiple alignment of a protein family as a training set in order to identify novel members of the family from their sequence data alone.

neuron A simple computational unit that performs a weighted sum on incoming signals, adds a threshold or bias term to this value to yield a net input, and maps this last value through an activation function to compute its own activation. Some neurons, such as those found in feedback or Hopfield networks, will retain a portion of their previous activation.

neutral mutation A mutation that does not change the fitness of the organism.

neutral theory The proposal that evolution at the molecular level is primarily determined by mutation and random genetic drift, rather than by natural selection.

New Chemical Entity (NCEs) Compounds identified as potential drugs that are sent from research and development into clinical trials to determine their suitability.

Newton's method An iterative method for finding the values at which a function evaluates to 0.

N-formylmethionine (fMet) The modified amino acid carried by the tRNA that is used during the initiation of translation in bacteria.

niche A way for an animal to make a living in an ecosystem.

niche overlap The extent to which two species require similar resources. It specifies the strength of the competition between the two species.

nick A position in a double-stranded DNA molecule where one of the polynucleotides is broken due to the absence of a phosphodiester bond.

nick translation A method of labelling DNA with radioactive or other modified nucleotides by using DNA polymerase I.

nitrogenous base A nitrogen-containing molecule, having the chemical properties of a base, that forms part of the molecular building blocks of DNA and RNA.

N-linked glycosylation The attachment of sugar units to an asparagine in a polypeptide.

node The graphical representation in a phylogenetic tree of an extant or ancestral operational taxonomic unit.

No Free Lunch (NFL) A theorem that states that in the worst case, and averaged over an infinite number of search spaces, all search methods perform equally well.

non-coding RNA An RNA molecule that does not code for a protein.

non-cooperative game A game in which pre-game agreements (e.g. to cooperate) are not enforceable.

nondegenerate site A nucleotide site in the coding region at which all substitutions are non-synonymous.

nondeterministic Permitting more than one choice of next move at some step in a computation.

Nondeterministic Finite Automaton (NFA) A finite state machine where for each pair of state and input symbols there may be several possible next states. This distinguished it from the DFA, where the next possible state is uniquely determined.

Nondeterministic Polynomial Time (NP) The complexity class of decision problems for which answers can be checked by an algorithm whose run time is polynomial in the size of the input.

non-disjunction The failure of homologous chromosomes to separate during meiosis.

non-functionalization The turning of a functional gene into a pseudogene following the occurrence of an incapacitating mutation.

nongenic DNA The portion of the genome that does not contain genes.

nonhomologous chromosomes These represent all the biological features of an organism and form a set, and the number of sets in a cell is called ploidy. In diploid organisms (most plants and animals), each member of a pair of homologous chromosomes is inherited from a different parent. But polyploid organisms have more than two homologous chromosomes, i.e., chromosomes that are not homologus.

Nonhomologous End-joining (NHEJ) Another name for the double-strand,break repair process.

non-linear A function that is not linear.

non-penetrance The situation whereby the effect of mutation is never observed during the lifetime of a mutant organism.

non-polar A hydrophobic chemical group.

non-redundant databases Researchers at the National

Center for Biotechnology Information (NCBI) coined the term "nr" database (non-redundant database) to refer to a database in which the obviously redundant entries have been merged. These entries are typically those that are 100%, character-by-character identical, and algorithms exist that can remove such redundancy. Although such a database has less redundancy than a primary database, a substantial amount of redundancy remains, and it can be removed only by a curator using scientific judgement.

nonsense mutation A point mutation in which a codon specific for an amino acid is converted into a "stop" codon.

nonsense strand The transcribed strand of a gene, the sequences of which is complementary to the RNA transcript.

non-synonymous mutation A mutation that alters a codon to that for another amino acid.

non-zero-sum game A game in which players have some common interests; a gain by one player does not require that others lose.

normal distribution A probability distribution, recognized as the traditional bell-shaped curve.

normal form A matrix representation of a simultaneous game.

normalization The process of structuring the schema of a relational database to eliminate or reduce ambiguity.

normalized library A cDNA library from which most of the highly expressed sequences have been removed in order to represent a greater proportion of low-abundance messenger RNAs.

northern blot An electrophoresis-based technique which is used to find mRNA sequences that are complementary to a piece of DNA called a probe.

northern blotting The transfer of RNA from an electrophoresis gel to a membrane prior to northern hybridization.

northern hybridization A technique used for detection of a specific RNA molecule against a background of many other RNA molecules.

not Recursively Enumerable (not-RE) An infinite set that cannot be recursively enumerated. Sets of this type that are also not co-recursively enumerable are effectively random.

NP-complete A subset of NP, the set of all decision problems whose solutions can be verified in polynomial time; problems that can be solved in polynomial time on a nondeterministic tuning machine.

NP-complete language A language in NP such that every language in NP can be reduced to it in polynomial time.

NP-hard A set or property of computational search problems. A problem is NP-hard if solving it in polynomial time would make it possible to solve all problems in class NP in polynomial time.

nuclear area In prokaryotic cells, a region containing the genetic information.

nuclear genome The DNA molecules present in the nucleus of a eukaryotic cell.

Nuclear Magnetic Resonance (NMR) A technique for determining the three-dimensional structure of large molecules.

nuclear matrix A proteinaceous scaffold-like network that permeates the cell.

nuclear pore An opening in the membrane of a cell's nuclear envelope that allows the exchange of materials between the nucleus and the cytoplasm.

nuclear pore complex The complex of proteins present at a nuclear pore.

nuclear receptor superfamily A family of receptor proteins that bind hormones as an intermediate step in modulation of genome activity by these hormones.

nuclear transfer A laboratory procedure in which a cell's nucleus is removed and placed into an oocyte with its own nucleus removed so the genetic information from the donor nucleus controls the resulting cell.

nuclease Any enzyme that can cleave the phosphodiester bonds of nucleic acid backbones.

nuclease protection experiment A technique that uses nuclease digestion to determine the positions of proteins on DNA or RNA molecules.

nucleic acid The term first used to describe the acidic chemical compound isolated from the nuclei of eukaryotic cells. Now used specifically to describe a polymeric molecule comprising nucleotide monomers, such as DNA and RNA.

nucleic acid hybridization Formation of a double-stranded hybrid by base-pairing between complementary polynucleotides.

nucleoid The DNA-containing region of a prokaryotic cell.

nucleolar organizing region   A part of the chromosome containing rRNA genes.

nucleolus   The region of the eukaryotic nucleus in which rRNA transcription occurs.

nucleoside   A five-carbon sugar covalently attached to a nitrogen base.

nucleosome   The spherical body formed by coils of chromatin; the assembly of histones and DNA that is the basic structural unit in chromatin.

nucleotide (nt)   A nucleic acid unit composed of a five carbon sugar joined to a phosphate group and a nitrogen base.

nucleotide diversity   A measure of polymorphism applied to nucleic acid sequences: the mean number of nucleotide differences per site between any two randomly chosen sequences from a population.

nucleotide excision repair   A repair process that corrects various types of DNA damage by excising and resynthesizing a region of a polynucleotide.

nucleotypic   A function of a DNA sequence other than as a carrier of genetic information (e.g. serving as a skeleton for the nucleus).

nucleus   The membrane-bound structure of an eukaryotic cell that contains the chromosomes.

null mutation   A mutation that leads to the absence of a gene product.

numerical solution   A solution to a problem that is calculated through a simulation.

O

Object Management Group (OMG) An open membership, not-for-profit consortium that produces and maintains computer industry specifications for interoperable enterprise applications.

Object-Oriented Programming (OOP) Computer languages and programming philosophy that emphasizes modularity among the elements of a program and their sharing of properties and intercommunication.

object-relational database Object databases combine the elements of object orientation and object-oriented programming languages with database capabilities. They provide more than persistent storage of programming language objects. Object databases extend the functionality of object programming languages (e.g. C++, Smalltalk, or Java) to provide full-featured database programming capability. The result is a high level of congruence between the data model for the application and the data model of the database. Object-relational databases are used in bioinformatics to map molecular biological objects (such as sequences, structures, maps and pathways) to their underlying representations (typically within the rows and columns of relational database tables.) This enables the user to deal with the biological objects in a more intuitive manner, as they would in the laboratory, without having to worry about the underlying data model of their representation.

obligate heterozygote An individual who is proved on the basis of pedigree information to carry one copy of a recessive allele even though it cannot be seen in his or her phenotype.

observed heterozygosity A measure of genetic variation in a population calculated as the mean frequency of heterozygotes over all loci.

occam's razor The notion that simpler explanations are generally better.

Okazaki fragment One of the short segments of RNA-primed DNA synthesized during replication of the lagging strand of the double helix.

oligogenic A phenotypic trait produced by two or more genes working together.

oligonucleotide array assay Oligonucleotide probes printed or synthesized onto a solid support to make a DNA chip.

oligonucleotide-directed mutagenesis An *in vitro* mutagenesis technique in which a synthetic oligonucleotide is used to introduce a predetermined nucleotide alteration into the gene to be mutated.

oligonucleotide (oligo) A short (2–50 nucleotides) sequence of (usually single-stranded) DNA which has been chemically synthesized for a specific experimental purpose.

oligonucleotide probe Short synthetic DNA sequence specifically designed to hybridize (attach) with the target DNA.

O-linked glycosylation The attachment of sugar units to a serine or threonine in a polypeptide.

oncogene A mutant gene that promotes uncontrolled cell growth once activated.

one-to-one A function or map that for every possible output has only one input that yields that particular output; if $f(a) = f(b)$, then $a = b$.

Online Analytical Processing (OLAP) Software for the real-time analysis of data stored in a database.

Online Mendelian Inheritance in Man (OMIM) Database of genetic diseases with information on their clinical diagnosis and treatment, cell biology, biochemistry, and molecular medicine.

Online Transaction Processing (OLTP) The processing of transactions by computers in real time, as opposed to in batch mode.

ontology In informatics an "ontology" is an organized collection of concepts.

Open Bioinformatics Foundation (OPEN-BIO) The purpose of the foundation is to act as an umbrella organization for the various bio*.org projects that grew out of the original BioPerl project. The goal of the foundation is to provide financial, administrative and technical assistance for our various open source life science projects.

Open Database Connectivity (ODBC) A framework for accessing and altering the contents of databases, developed by Microsoft Corporation.

open-loop control Partially automated control method in which a part of the control system is given over to humans.

open promoter complex A structure formed during assembly of the transcription initiation complex consisting of the RNA polymerase and/or accessory proteins attached to the promoter, after the DNA has been opened up by breakage of base pairs.

Open Reading Frame (ORF) A series of codons starting with an initiation codon and ending with a termination codon. It is the part of a protein-coding gene that is translated into protein.

open system Computer industry term for computer hardware and software that is built to common, public standards, allowing purchasers to select components from a variety of vendors and use them together.

Operational Taxonomic Unit (OTU) One of the organisms being compared in a phylogenetic analysis.

operator The nucleotide sequence to which a repressor protein binds to prevent transcription of a gene or operon.

operon A set of one or more structural genes, an operator and a promoter that are transcribed as a unit and are expressed in a coordinated manner; they regulate bacterial gene expression.

optical mapping A technique for the direct visual examination of restricted DNA molecules.

optimal alignment An alignment of two sequences with the highest possible score.

optimization The process of finding the parameters that minimize or maximize a function.

organelle A functional membrane-enclosed structure within eukaryotic cells, such as mitochondria and chloroplasts.

organism An individual, composed of organ systems (if multicellular). Multiple organisms make up a population.

origin of replication A site on a DNA molecule where replication initiates.

Origin Recognition Complex (ORC) A set of proteins that binds to the origin recognition sequence.

origin recognition sequence A component of a yeast origin of replication.

orphan family A group of homologous genes whose functions are unknown.

Orthogonal Field Alternation Gel Electrophoresis (OFAGE) An electrophoresis system in which the field alternates between pairs of electrodes set at an angle of 45°, used to separate large DNA molecules.

ortholog Genes in different species that evolved from a common ancestral gene by speciation. Normally, orthologs retain the same function in the course of evolution. Identification of orthologs is critical for reliable prediction of gene function in newly sequenced genomes.

orthologous Homologous sequences in different species that arose from a common ancestral gene during speciation, that may or may not be responsible for a similar function.

orthology Sequence similarity as a consequence of a speciation event.

outcome The consequence for a player of a specific combination of all player's strategies.

out-degree The number of edges in a directed graph that begin at (point from) the specified vertex.

outer product An operation on two vectors that yields a matrix.

outgroup A species or a set of species that is the least related to the others in a group of species.

"Out of Africa" Hypothesis A hypothesis that holds the modern human populations (*Homo sapiens*) are all derived from a single speciation event that took place in a restricted region in Africa.

out-of-phase overlapping The encoding of two or more proteins in different frames of the same DNA sequence.

overdominance A selection regime resulting from the heterozygote having a higher fitness than either of the homozygotes.

overlap A part of a sequence that is common to sequences derived from separate sequencing experiments.

overlapping clones Collection of cloned sequences made by generating randomly overlapping DNA fragments with infrequently cutting restriction enzymes.

overlapping genes Two genes whose coding regions overlap.

ovum The female gamete, egg.

P

p53 A gene which normally regulates the cell cycle and protects the cell from damage to its genome.

Pacific Symposium on Bio-computing (PSB) "An international, multidisciplinary conference for the presentation and discussion of current research in the theory and application of computational methods in problems of biological significance."

packet-switched networks Computer networks which exchange the messages between different computers not by seizing a dedicated circuit, but by sending a message in a number of uniformly sized packets along common channels, shared with other computers.

pair-rule genes Developmental genes that establish the basic segmentation pattern of the *Drosophila* embryo.

pairwisealignment Inthefirststep, two sequences are padded by gaps so that they are the same length and so that they display the maximum similarity on a residue-to-residue basis in the first step. An alignment which has the maximum amount of similarity with the minimum number of residue "substitutions".

palindrome A region of DNA with a symmetrical arrangement of bases occuring about a single point such that the base sequences on either side of that point are identical (if the strands are both read in the same direction) e.g. 5´ GAATTC 3´ whose complementary sequence is 3´ CTTAAG 5´.

PAM matrix PAM (Per cent Accepted Mutation) and BLOSUM (Blocks Substitution Matrix) are the matrices that define scores for each of the 210 possible amino acid substitutions.

pan-editing The extensive insertion of nucleotides into an abbreviated RNA, resulting in a functional molecule.

paracentric inversion An inversion does not include the centromere.

parallel In computing, it refers to processes that occur simultaneously.

parallel evolution The development of similar characteristics in organisms that are not closely related (not part of a monophyletic group) due to adaptation to similar environments and/or strategies of life.

parallel substitution The independent occurrence of the same mutation at the same nucleotide site in two or more lineages.

paralog The genes which are related by duplication within a genome. Orthologs retain the same function in the course of evolution, whereas paralogs evolve new functions, even if these are related to the original one.

paralogous Homologous sequences (sequences that share a common evolutionary ancestor) that diverged by gene duplication, as opposed to orthologs, which diverged by speciation.

paralogy Sequence similarity between the descendants of a duplicated ancestral gene.

parameters The user-selectable values, typically experimentally determined, that govern the boundaries of an algorithm or program. For instance, selection of the appropriate input parameters govern the success of a search algorithm. Some of the most common search parameters in bioinformatics tools include the stringency of an alignment search tool, and the weights (penalties) provided for mismatches and gaps.

paranemic Refers to a helix whose strands can be separated without unwinding.

pararetrovirus A virus that contains a gene for reverse transcriptase but cannot insert itself into the host chromosome.

parasite An organism that lives in, with, or on another organism. The parasites benefit from the association without contributing to the host, usually they cause some harm to the host.

parasitism A form of symbiosis in which the population of one species benefits at the expense of the population of another species.

parse The process of analysing a sequence of tokens (e.g. words) to determine their grammatical structure with respect to a given (more or less) formal grammar. Flat file data often have to be parsed into separate files prior to analysis.

parsimony The use of a minimum number of means to achieve an end.

partial linkage A type of linkage usually displayed by a pair of genetic and/or physical markers on the same chromosome, the markers not always being inherited together because of the possibility of recombination between them.

partial restriction Digestion of DNA with a restriction endonuclease under limiting conditions so that not all restriction sites are cut.

partly open barrel The edge strands of residues are not properly hydrogen bonded because one of the strands is in two parts connected with a linker of more than one residue.

patent In genetics, it refers to conferring the right or title to genes, gene variations, or identifiable portions of sequenced genetic material to an individual or organization.

path A set of distinct vertices of a graph, in which each vertex is connected by an edge to the next.

path graph A figure showing the locations of the pointers in a dynamic programming alignment.

pathways Bioinformatics strives to define representations of key biological datatypes, algorithms and inference procedures, including sequences, structures, biological pathways and reactions. Representing and computing with biological pathways requires ontologies for representing pathway knowledge; User interfaces to these databases; Physico-chemical properties of enzymes and their substrates in pathways; And pathway analysis of whole genomes including identifying common patterns across species and species differences.

pathway profiling The concurrent activity of a set of genes or proteins.

pattern A pattern language must be defined in order to apply different criteria to different positions of a sequence. In order to have position-specific comparison done by a computer, a pattern-matching algorithm must allow alternative residues at a given position, repetitions of a residue, exclusion of alternative residues, weighting, and ideally, combinatorial representation. Molecular biological patterns usually occur at the level of the characters making up the gene or protein sequence.

pattern classification A task that neural networks which are often trained to do.

pattern of substitution The relative frequency with which a nucleotide or an amino acid changes into another during evolution.

PCR (polymerase chain reaction; *in vitro* **DNA amplification)** The laboratory technique for duplicating (or replicating) the DNA using the bacterium *Thermus aquaticus*, a heat stable bacterium from the hot springs of Yellowstone.

P1-derived Artificial Chromosome (PAC) A vector used to clone DNA fragments (100 to 300 kb insert size; average, 150 kb) in *Escherichia coli* cells.

peano curve A fractal space-filling curve that can fill a plane even though it is a line of infinite length.

pedigree The genetic relationships between the members of a family, often with information on the inheritance of one or more conditions or genetic loci.

pedigree analysis The use of pedigree charts to analyse the inheritance of a genetic or DNA marker in a human family.

P-element A DNA transposon of *Drosophila*.

penetrance A defect or change in a gene that is common to a species will lead to a trait or disease.

pentose A sugar with five carbon atoms, such as ribose or deoxyribose.

peptide A short stretch of amino acids; each amino acids are covalently coupled by a peptide (amide) bond.

peptide bond A covalent bond formed between two amino acids when the amino group of one is linked to the carboxyl group of another (resulting in the elimination of one water molecule).

Peptide Nucleic Acid (PNA) A polynucleotide analog in which the sugar-phosphate backbone is replaced by amide bonds.

peptidyl site The site in the ribosome occupied by the tRNA attached to the growing polypeptide during translation.

peptidyl transferase An enzyme that synthesizes peptide bonds during translation.

perceptron The simplest type of feed-forward neural network. It has only inputs and outputs, i.e., no hidden layers.

pericentric inversion An inversion that includes the centromere.

periodic Refers to motion that goes through a finite number of regions, returns to a previous state, and repeats the same fixed pattern forever.

Perl (Practical Extraction and Reporting Language) An interpreted computer language for easily manipulating texts, files and processes.

permissive conditions Conditions under which a conditional-lethal mutant is able to survive.

permutation One of the possible arrangements of a particular set of characters.

perturbation A small nudge or displacement of a dynamical system from its current state, either by an external force or random variation.

PEST sequences Amino acid sequences that influence the degradation of proteins in which they are found.

phage (bacteriophage) A virus that infects bacterial cells and serves as a useful vector for introducing genes into bacteria for a number of purposes.

phage display A technique in which phages are engineered to fuse a foreign peptide or protein with their capsid (surface) proteins and hence display it on their cell surfaces. The immobilized phage may then be used as a screen to see what ligands bind to the expressed fusion protein exhibited (displayed) on the phage surface.

pharmaceutical bioinformatics Bioinformatics and structure-aided drug design are really part of the same continuum. Bioinformatics offers a means to get a structure through sequence; while structure-aided drug design offers a drug through structure.

pharmacogenomics A method that employs genetic and genomic information to predict the response of individual patients and patient populations to drugs.

pharmacology The science of the action of drugs and other chemicals on living biological systems.

pharmacophore The three-dimensional spatial arrangment of atoms, substituents, functional groups, or chemical features that together are sufficient to describe the pharmacologically active components of a drug molecule or molecule series.

phase class The position of an intron relative to the reading frame of the two adjacent protein-coding exons.

phase-0 intron   An intron that lies between two codons.

phase-1 intron   An intron that lies between the first and second nucleotides of a codon.

phase-2 intron   An intron that lies between the second and third nucleotides of a codon.

phase transition   In dynamical systems, a change from one mode of behaviour to another, such as from continuous to chaotic behaviour.

phenetics   A classification approach based on the numerical typing of as many phenotypic characters as possible.

phenocopy   An individual in whom a phenotype, normally associated with a mutant gene, has been brought about by some factor in the environment.

phenogram   A graphical representation that portrays or attempts to portray the taxonomic relationships among a number of individuals, species, or higher taxa on the basis of overall similarities between them.

phenotype   Any observable feature of an organism that is the result of one or more genes.

phenylalanine (Phe, F)   An α-amino acid with the formula $HO_2CCH\,(NH_2)CH_2C_6H_5$, which is found naturally in the breast milk of mammals and manufactured as food and drink products and are also sold as nutritional supplements for their reputed analgesic and antidepressant effects. Phenylalanine is structurally closely related to dopamine, epinephrine (adrenaline) and tyrosine.

phosphate group   A chemical group composed of a central phosphorous bonded to three or four oxygens. The net charge of the group is negative.

phosphodiesterase   A type of nuclease.

phosphodiester bond   The chemical link between adjacent nucleotides in a polynucleotide.

phosphorimaging   An electronic method for determining the positions of radioactive markers in a microarray or on a hybridization membrane.

phosphorylation   The chemical attachment of phosphorous to a molecule, usually associated with the storage of energy in the covalent bond that is also formed.

photolyase   *See* DNA photolyase.

photoproduct A modified nucleotide resulting from treatment of DNA with UV radiation.

photoreactivation A DNA repair process in which cyclobutyl dimers and (6-4) photoproducts are corrected by a light-activated enzyme.

PHP Hypertext Preprocessor A server-side, cross-platform, HTML-embedded scripting language that lets web developers create dynamic content that interacts with databases.

phrap Developed by Phil Green at the University of Washington, "PHil's Revised Assembly Program" is a tool for assembling shotgun-sequenced DNA fragments.

PHYLIP A program package created by J. Felsenstein for phylogenicity.

phylogenetics The reconstruction of the evolutionary history of a group of taxa or genes.

phylogenetic tree The graphical representation of the phylogeny of a group of taxa or genes.

phylogeny A classification scheme that indicates the evolutionary relationships between organisms.

phylum The broadest taxonomic category within kingdoms.

physical map A map of the locations of identifiable landmarks on DNA (e.g. restriction enzyme cutting sites, genes), regardless of inheritance.

physical mapping The use of molecular biology techniques to construct a genome map.

pilus A structure involved in bringing a pair of bacteria together during conjugation; possibly the tube through which DNA is transferred.

PIMA An alignment algorithm developed by Smith and Smith which performs multiple alignments using a covering pattern construction algorithm.

pi–pi interactions The hydrophobic interactions that occur between adjacent base pairs in a double-stranded DNA molecule.

planning In computer science, and particularly in artificial intelligence, the task of determining a stepwise plan to accomplish a very specific task.

plaque A zone of clearance on a lawn of bacteria caused

by lysis of the cells by infecting bacteriophages.

plasmid Any replicating DNA element that can exist in the cell independently of the chromosomes. Synthetic plasmids are used for DNA cloning and it is commonly found in bacterial cells.

plectonemic Refers to a double helix whose strands cannot be separated without unwinding.

pleiotropic A term describing a genotype with multiple phenotypic effects.

pleiotropy Multiple, apparently unrelated phenotypic effects of mutation at a single gene.

plesiomorphic character state A character state possessed by a remote common ancestor of a group of organisms.

pluripotency The potential of a cell to develop into more than one type of mature cell, depending on environment.

P-N-glycosidic bond The chemical bond between the base and sugar of a nucleotide.

POD domain Plain Old Documentation. A DNA-binding motif found in a variety of proteins.

point mutation A mutation in which a single nucleotide in a DNA sequence is substituted by another nucleotide.

polar A hydrophilic (water-loving) chemical group.

polarity The property of nucleic acids to be read only one way from 5′ to 3′ and differently in the opposite direction.

pole One of the two ends of a spindle where the microtubules converge and to which the chromosomes will eventually be pulled.

Polyacrylamide Gel Electrophoresis (PAGE) Electrophoresis carried out in a polyacrylamide gel and used to separate DNA molecules between 10 and 1500 bp in length.

polyadenylate-binding protein A protein that aids to poly(A) polymerase during polyadenylation of eukaryotic mRNAs, and which plays a role in maintenance of the tail after synthesis.

polyadenylation The post-transcriptional addition of a series of adenines to the 3′ end of an eukaryotic mRNA.

polyadenylation editing A form of editing that occurs with many animal mitochondrial RNAs, resulting in a termination codon being formed by adding a poly(A) tail to an mRNA that ends with the nucleotides U or UA.

polyadenylation signal A sequence region on most eukaryotic mRNA molecules that specifies the location of the polyadenylation site.

polyadenylation site The 3´ end of most mRNA molecules in eukaryotes, at which the mRNA has been cleaved and a tail of 200–300 adenosine nucleotides (the poly(A) tail) has been added.

poly(A) polymerase An enzyme that attaches a poly(A) tail to the 3´ end of an eukaryotic mRNA.

poly(A) tail The stretch of adenine (A) residues at the 3´ end of eukaryotic mRNA that is added to the pre-mRNA as it is processed, before its transport from the nucleus to the cytoplasm and subsequent translation at the ribosome.

polyclonal A colony of cells made up of more than one clone. In immunology, a group of cells expressing a variety of different antibodies.

polygamy A mating system in which a male mates with more than one female (polygyny) or a female mates with more male (polyandry).

polygenic disorder A genetic disorder resulting from the combined action of alleles of more than one gene (e.g. heart disease, diabetes, and some cancers).

polygenic inheritance The inheritance of a characteristic which is determined by the cumulative actions of many different genes, each with small individual effects.

polymer A compound made up of a long chain of identical or similar units.

polymerase An enzyme that catalyses the covalent joining of nucleotides, e.g. DNA polymerase and RNA polymerase.

polymorphic Refers to a locus that is represented by a number of different alleles or haplotypes in the population as a whole.

polymorphic marker A length of DNA that displays population-based variability so that its inheritance can be followed.

polymorphism Difference in DNA sequence among individuals of a given species due to the presence of two or more alleles of a given locus.

polynomial function A function in which the output is the sum of terms that are the products of constant values and the input raised to some integer power.

polynomial time The class of decision problems that can be answered within a number of computational steps that is a polynomial function of the size of the problem.

polynucleotide A single-stranded DNA or RNA molecule.

polynucleotide kinase An enzyme that adds phosphate groups to the 5′ ends of DNA molecules.

polypeptide A linear chain of amino acids joined head to tail via a peptide bond between the carboxylic acid group of one amino acid and the amino group of the next amino acid.

polyphyletic Descended from different ancestors.

polyploidy It possesses more than the normal two haploid sets of chromosomes.

polyprotein A translation product consisting of a series of linked proteins that are processed by proteolytic cleavage to release the mature proteins.

polypyrimidine tract A pyrimidine-rich region near the 3′ end of a GU–AG intron.

polysome An mRNA molecule that is being translated by more than one ribosome at the same time.

population A group of individuals in a species that share a common gene pool.

population genetics The study of variation in genes among a group of individuals.

positional cloning A technique used to identify genes, usually those associated with diseases, based on their location on a chromosome. A procedure that uses information on the map position of a gene to obtain a clone of that gene.

positive feedback Biochemical control where the accumulation of the product stimulates the production of an enzyme which is responsible for that product's production.

positive feedback control Occurs when information produced by the feedback increases and accelerates the response.

positive selection Selection for an advantageous mutant allele.

posterior probability The probability of an event after evidence has been considered.

post-production system A model of computation that resembles a collection of "if ... then" rules.

post-Replication Complex (post-RC) A complex of proteins, derived from a pre-replication complex, that forms at a eukaryotic origin of replication during the replication process and ensures that the genome is copied just once per cell cycle.

post-translational modification Alterations made to a protein after its synthesis at the ribosome. These modifications, such as the addition of carbohydrate or fatty acid chains, may be critical to the function of the protein.

POU domain A DNA-binding motif found in a variety of proteins.

predator-prey system An ecosystem in which one portion of the population consumes another.

predicate A function whose output is either "true" (1) or "false" (0).

prefix A substring at the beginning of a given string.

preinitiation complex The structure comprising the small subunit of the ribosome, the initiator tRNA plus ancillary factors that form the initial association with an mRNA during protein synthesis.

pre-messenger RNA The primary transcript of a protein-coding gene before maturation.

pre-mRNA The primary transcript of a protein-coding gene.

prepriming complex A complex of proteins formed during initiation of replication in bacteria.

preproprotein The primary product of translation before any post-translational changes have been made.

pre-Replication Complex (pre-RC) A protein complex in a eukaryote that is constructed at an origin of replication and enables replication to begin.

pre-RNA The initial product of transcription of a gene or group of genes that is subsequently processed to give the mature transcript(s).

pre-rRNA The primary transcript of a gene or group of genes that specify rRNA molecules.

presymptomatic screening Testing to detect genetic disorders that only become apparent later in life.

pretermination codon A codon that requires only one mutation to become a termination codon.

pre-tRNA The primary transcript of a gene or group of genes that specify tRNA molecules.

pribnow box A component of the bacterial promoter, i.e., a base sequence in prokaryotic promoters to which RNA polymerase binds in an early step of initiating transcription.

primary amino acid One of the 20 amino acids specified by the universal genetic code.

primary key A key that uniquely identifies each record in a relational table.

primary sequence (protein) The linear sequence of a polypeptide or protein.

primary structure The sequence of amino acids in a polypeptide.

primary transcript The initial product of transcription of a gene or group of genes, which is subsequently processed to give the mature transcript(s).

primase The RNA polymerase enzyme that synthesizes RNA primers during bacterial DNA replication.

prime number A natural number that can be evenly divided only by itself and 1.

primer A short oligonucleotide that is attached to a single-stranded DNA molecule in order to provide a starting point to which new deoxy-ribonucleotides can be added by DNA polymerase.

primosome A protein complex involved in DNA replication.

prim's algorithm An algorithm for computing a minimum spanning tree.

Principal Component Analysis (PCA) A procedure that attempts to identify patterns in a large dataset of variable character states.

principle of independent assortment During gamete formation, alleles in one gene pair segregate into gametes independently of the alleles of other gene pairs.

principle of segregation Each pair of factors of heredity separate during gamete formation so that each gamete receives one member of a pair.

PRINTS The protein motif fingerprint database created by Attwood and Beck.

prion An unusual infectious agent that consists of pure protein.

prior probability The probability of an event before evidence is considered. Often, the prevalence of a disorder in a population.

Prisoner's Dilemma (PD) A non-zero-sum game in which both players have incentive not to cooperate under any circumstances.

privacy In genetics, the right of people to restrict access to their genetic information.

probability The expectation of the occurrence of a particular event; the likelihood that a random event will occur.

proband The affected member in a pedigree through whom the family came to medical or scientific attention.

probe A single-stranded DNA or RNA molecule with a specific base sequence, labelled either radioactively or immunologically, that is used to detect the complementary base sequence by hybridization.

processed pseudogene A pseudogene that results from integration of an mRNA into the genome by reverse transcriptions.

processivity The length of polynucleotide that is synthesized by a DNA polymerase before it dissociates from the template.

producer The first level in a food pyramid; an organism that generates the food used by all other organisms in the ecosystem.

profile A table that lists the frequencies of each amino acid in each position of protein sequence. Frequencies are calculated from multiple alignments of sequences containing a domain of interest.

program An algorithm that is written in a programming language for execution on a physical computer.

programmed mutation The possibility that under some circumstances an organism can increase the rate at which mutations occur in a specific gene.

Programs for Genomic Applications (PGA) A major initiative of the U.S. National Heart, Lung, and Blood Institute (NHLBI) to advance functional genomic research related to heart, lung, blood, and sleep health and disorders.

progressive alignment A multiple alignment algorithm in which the sequences are first clustered and then added one by one, in order of decreasing similarity, to the growing multiple alignment.

prokaryote An organism whose cells lack a membrane-bound, structurally discrete nucleus and other subcellular compartments.

Proliferating Cell Nuclear Antigen (PCNA) An accessory protein involved in DNA replication in eukaryotes.

proline (Pro, P) An α-amino acid, one of the twenty DNA-encoded amino acids. Its codons are CCU, CCC, CCA and CCG.

proline-rich domain A type of activation domain.

promiscuous DNA DNA that has been transferred from genome of one organelle to another.

promoter The short sequence of nucleotides on DNA that start the transcription of RNA by RNA polymerase.

promoter clearance The completion of successful initiation of transcription that occurs when the RNA polymerase moves away from the promoter sequence.

pronucleus The nucleus of a sperm or an egg prior to fertilization.

proof A sequence of statements in which each subsequent statement is derivable from one of the previous statements or from an axiom of a formal system. The final statement of a proof is usually the theorem that one has set out to prove.

proofreading The $3' - 5'$ exonuclease activity possessed by some DNA polymerases which enables the enzyme to replace a misincorporated nucleotide.

propensity In the Chou and Fasman secondary structure prediction, the measure of a residue's tendency to assume a given structure.

prophage The integrated form of the genome of a lysogenic bacteriophage.

prophase A stage at the beginning of mitosis or meiosis during which the chromosomes condense and become visible.

proprietary databases Fee-based, copyrighted databases (in contrast to public databases such as those at DDBJ/ EMBL/ GenBank).

proprotein A product of translation after the signal peptide has been removed and before additional post-translational modifications have been made.

PROSITE A database of "patterns" (regular expressions) specific for various protein motifs.

prosthetic group A nonprotein molecule attached to an apoprotein that is required for functionality (e.g. haem in haemoglobin).

protease An enzyme that breaks down protein.

proteasome A multisubunit protein structure that is involved in the degradation of other proteins.

protein A molecule composed of one or more chains of amino acids in a specific order; the order is determined by the base sequence of nucleotides in the gene coding for the protein.

protein bioinformatics Bioinformatics and experimental analysis of protein superfamilies for understanding protein structure–function relationships and developing strategies for protein engineering. Using superfamily analysis to understand how protein sequence and structure determine protein function. Our computational approach begins with identifying the sets of divergently related proteins that comprise enzyme superfamilies and then attempts to correlate their conserved and variable structural

features to similarities and differences in their functions.

Protein Data Bank (PDB) A database for the processing and distribution of 3D biological macromolecular structure data, maintained by the Research Collaboratory for Structural Bioinformatics (RCSB). An international repository for the results of macromolecular studies using NMR, X-ray crystallography, or homology methods. The results of structural studies of proteins, RNA, DNA, viruses, and polysaccharides are presently available. The term PDB also defines a standard file format for publishing protein and nucleotide structures for use in computer programs.

protein engineering Various techniques for making directed alterations in protein molecules, often to improve the properties of enzymes used in industrial processes.

protein expression profiling The pattern of expression of a protein.

Protein Extensible Markup Language (PROXIML) A standard programming language for describing protein data.

protein family A set of proteins that share a common evolutionary origin reflected by their relatedness in function, which is usually demonstrated by similarities in sequence, or in primary, secondary, or tertiary structure.

protein folding The adoption of a folded structure by a polypeptide.

Protein-fragment Complementation Assays (PCA) A process that facilitates the detection of protein–protein complexes following the expression of full-length mammalian genes linked in-frame to polypeptide fragments of rationally dissected reporter genes. If cellular activity causes the association of two proteins linked to complementary reporter fragments, the interaction of the proteins of interest enables refolding of the fragments, which then generates a fluorescent signal.

Protein Information Resource (PIR) An integrated public bioinformatics resource that supports genomic and proteomic research and scientific studies.

protein-coding gene A gene that contains a reading frame, the mRNA which is translated.

proteinoid A polymer of amino acids formed spontaneously from inorganic molecules; has enzyme-like properties and can catalyse chemical reactions.

protein–protein cross-linking A technique that links together adjacent proteins in order to identify proteins that are positioned close to one another in a structure such as a ribosome.

proteome The complete profile of proteins expressed in a given tissue, cell or biological system at a given time.

proteomics The study of protein expression, structure and function, and the interactions of all proteins of a specific organism.

protogenome An RNA genome that existed during the RNA world.

proto-oncogene A normal gene that can be converted to an oncogene by one of the several mechanisms.

prototroph An organism that has no nutritional requirements beyond those of the wild type and which can grow on minimal medium.

provirus A viral genome integrated into the genome of the host cell.

pseudogene An inactivated and hence non-functional copy of a gene.

pseudograph A graph that contains a vertex connected to itself by an edge, forming a loop.

pseudoknot A 3D structure where the RNA folds back on a hairpin to form a structure where three strands are held together by hydrogen bonds.

PSI-BLAST (Position-Specific Iterative BLAST) An iterative search using the BLAST algorithm. A profile is built after the initial search, which is then used in subsequent searches. The process may be repeated, if desired with new sequences found in each cycle used to refine the profile.

PSSM (Position-Specific Scoring Matrix) The PSSM gives the log-odds score for finding a particular matching amino acid in a target sequence.

PubEST Abbreviation for a sequence from a public-domain source, such as the WashU-Merck EST Project or Banting Institute.

public databases Freely accessible databases such as GenBank/ EMBL/ DDBJ, Array-Express or BLOCKS. There has been much debate about public vs proprietary databases.

PubMed A portal for retrieval of bibliographic information from MedLine and other sources via the world wide web.

pulsed-field gel electrophoresis A technique of elec-trophoresis which is able to resolve very large DNA molecules by periodic alterations of the direction of the applied electric field.

punctuated equilibrium The theory states that evolution is characterized by geologically long periods of stability during which little speciation occurs, punctuated by short periods of rapid change.

punctuation codon A codon that specifies either the sex or the end of a gene.

Punnett square A tabular analysis for predicting the genotypes of the progeny resulting from a genetic cross.

purifying selection A selection regime resulting in the removal of an allele from the population.

purine A nitrogen-containing, double-ring, basic compound that occurs in nucleic acids. The purines in DNA and RNA are adenine and guanine.

PushDown Automaton (PDA) A finite state machine that incorporates an unlimited "stack" on which symbols can be stored (pushed) and retrieved (popped). PDAs recognize context-free languages.

putative Commonly accepted or supposed.

P-value The probability of an alignment occurring with the score in question or better. The P-value is calculated by relating the observed alignment score, S, to the expected distribution of HSP scores from comparisons of random sequences of the same length and composition as the query to the database. The most highly significant P-values will be those close to 0. P-values and E-values are different ways of representing the significance of the alignment.

pyrimidine A nitrogen-containing, single-ring, basic compound that occurs in nucleic acids. The pyrimidines in DNA are cytosine and thymine; in RNA, cytosine and uracil.

pyrosequencing A novel DNA- sequencing method in which addition of a nucleotide to the end of a growing polynucleotide is directly detected by conversion of the released pyrophosphate into a flash of chemluminescence.

Python A computer language, similar to Perl, used for easily manipulating texts, files and processes.

Q

qualitative reasoning A sub-speciality of artificial intelligence concerned with inference and knowledge representation when knowledge is not precisely defined.

quantitative PCR A PCR method that enables the number of DNA molecules in a sample to be estimated.

Quantitative Structure Activity Relationship (QSAR) Relates numerical properties of the molecular structure to the activity via a mathematical model.

Quantitative Trait Locus (QTL) A gene locus that contributes to a complex trait.

quantum model of speciation A model of evolution that holds that speciation sometimes occurs rapidly as well as over long periods, as the classical theory proposed.

quasiperiodic A form of motion that is regular but never exactly repeating.

quaternary structure The structure formed by association of two or more polypeptides.

query The input sequence (or other type of search term) with which all of the entries in a database are to be compared.

R

radiation hybrid A hybrid cell produced by the fusion of a human cell and a rodent cell, after the human cell has been irradiated with gamma rays to fragment the DNA.

radiation hybrid mapping The approach to physically mapping DNA that makes use of the frequency of X-ray induced breakage to infer distances between markers.

radiation hybrid (RH) panel A set of DNA samples prepared from a collection of radiation hybrids.

radical substitution The substitution of an amino acid by another with markedly different chemical properties.

radioactive marker A radioactive atom incorporated into a molecule and whose radioactive emissions are subsequently used to detect and follow that molecule during a biochemical reaction.

radiolabelling The technique for attaching a radioactive atom to a molecule.

Ramachandran plot A contour plot of the different phi and psi angles that are found within a protein.

random Without evident cause; not compressible; obeying the statistics of a fair coin toss.

random genetic drift The random process that leads to gradual changes in the frequency of alleles in a population from one generation to the next.

random mating Selection of a mate without considering their genotype or any trait associated with a particular genotype.

random walk A walk in one or more dimensions that is dictated by the outcome of some random source.

Rapid Amplification of cDNA Ends (RACE) A PCR-based technique for mapping the end of an RNA molecule.

ras A protein involved in signal transduction.

RASMOL Program package by R. Sayle to display protein structures.

rate of gene substitution The number of gene substitutions per locus per unit time.

rate of mutation The number of mutations per locus or nucleotide site per unit time, usually per generation time.

Rat Genome Database (RGD) A collaborative effort among leading research institutions involved in rat genetic and genomic research to collect, consolidate, and integrate data generated from ongoing rat genetic and genomic research efforts and make these data widely available to the scientific community.

rate of nucleotide substitution The number of nucleotide substitutions per nucleotide site per unit time.

rational drug design The development of drugs based on the 3-dimensional molecular structure of a particular target.

rational number A real number that can be expressed as a fraction.

rawscore(S) Thescoreofanalignment6666666666666666666, S, calculated as the sum of substitution and gap scores.

reading frame A sequence of codons beginning with an initiation codon and ending with a termination codon; also the way in which nucleotides are read in groups of three (codons) to specify the polypeptide coded by a gene.

reading frame (also open reading frame) The stretch of triplet sequence of DNA that encodes a protein. The reading frame is designated by the initiation or start codon and is terminated by a stop codon.

read-through mutation A mutation that changes a termination codon into a codon specifying an amino acid, and hence results in read-through of the termination codon.

reagent A source of biological or chemical material that can be used as a starting block in laboratory experiments.

real number A number that has no imaginary part; comprises the rational numbers and the irrational numbers.

RecA A protein that is involved in homologous recombination in *E. coli*.

recall   A measure of information retrieval.

RecBCD enzyme   An enzyme complex involved in homologous recombination in *E. coli*.

recessive   An allele that is not expressed in a heterozygote.

reciprocal strand exchange   The exchange of DNA between two double-stranded molecules, occurring as a result of recombination, such that the end of one molecule is exchanged for the end of the other molecule.

reciprocal translocation   A chromosomal rearrangement in which material is exchanged between two nonhomologous chromosomes.

recognition helix   An alpha-helix in a DNA-binding protein, that is responsible for recognition of the target nucleotide sequence.

recognition sequence   The sequence recognized by a restriction endonuclease; in many cases, a short palindrome.

recombinant   An offering that possesses neither of the combinations of alleles displayed by the parents. A chromosome (or individual) arising from a crossover at meiosis.

recombinant clone   A clone containing recombinant DNA molecules.

recombinant DNA (rDNA) A form of DNA produced by splicing together segments of DNA from two or more organisms.

recombinant DNA technology   The procedures used to join together DNA segments in a cell-free system (an environment outside a cell or organism) to make a recombinant DNA. The techniques involved in the construction, study and use of recombinant DNA molecules.

recombinant protein   A protein synthesized in a recombinant cell due to expression of a cloned gene.

recombinase   An enzyme that catalyses site-specific recombination events.

recombination   The exchange of regions of DNA on chromosomes via crossover during meiosis.

recombination frequency   The proportion of recombinant progeny arising from a genetic cross.

recombination hotspot   A region of a chromosome where crossovers occur at a higher frequency than the average for the chromosome as a whole.

recombination repair   A repair process that mends double-stranded breaks in a DNA.

record In a relational database, each record corresponds to a row (tuple) in a table.

recurrence risk The chance that a genetic disease may occur in siblings (or other relatives) of an affected child.

recurrent neural network A network similar to a feedforward neural network except that there may be connections from an output or hidden layer to the inputs. Recurrent neural networks are capable of universal computation.

recursion The process of solving a large problem by reducing it to one or more sub-problems, which are solved in similar fashion.

recursive A set of function is recursive if it is computable.

Recursively Enumerable (RE) A potentially infinite set whose members can be enumerated by a universal computer (Turing machine).

red queen hypothesis The hypothesis of co-evolution that organisms must keep evolving simply to maintain themselves in constantly changing environments.

reduction division The first meiotic division in which the chromosome number is reduced by half.

reductionism The idea that nature can be understood by dissection.

redundancy The presence of more than one identical item.

redundant databases When sequence databanks were first created, primary (redundant) databases had the advantage of being more comprehensive than curated databases and more likely to contain recently discovered sequences. However, redundancy is no longer much of an advantage. In a highly redundant database, biologically significant results are more likely to be hidden among large numbers of irrelevant reported matches.

referential integrity A database concept that asserts a value appearing in one context also appears in another, related context.

regional duplication A duplication involving less than the entire genome.

Regular Expression (RE) A string of characters specifying a pattern. Usually, wildcards (match to anything) and variable positions (match to one of the following) are allowed.

Regular Language (RL) A set, or language, that is accepted by a finite automation.

regulator gene A gene that prevents or represses the activity of the structural genes in an operon.

regulatory gene A non-transcribed gene; sometimes used to denote a structural gene engaged in the regulation of gene expression.

regulatory mutant A mutant that has a defect in a promoter or other regulatory sequence.

regulatory region The portion of a DNA sequence that controls expression of a particular gene.

relation A predefined row/column format for storing information in a relational database.

relational database A database that allows the definition of data structures, storage and retrieval operations and integrity constraints. Data and relations between them are organized in tables.

Relational Database Management Systems (RDBMS) A software system that includes a database architecture, query language, and data loading and updating tools and other ancillary software that together allow the creation of a relational database application.

relative rate test A calibration-free test for checking the constancy of the rate of nucleotide substitutions in different lineages during their evolution, thus determining whether or not the molecular clock operates at the same rate among different lineages.

release factor A protein that plays an ancillary role during termination of translation.

renaturation The return of a denatured molecule to its natural state. Restoration of the normal three-dimensional structure of a macromolecule; in reference to nucleic acids, the term means the formation of a double-stranded molecule by complementary base pairing between two single-stranded molecules.

repeats (repeat sequences) Repeat sequences and approximate repeats occur throughout the DNA of higher organisms (mammals).

repetitive DNA A sequence that is repeated two or more times in a DNA molecule or genome.

repetitive DNA fingerprinting A clone fingerprinting technique that involves determining the positions of genome-wide repeats in cloned DNA fragments.

repetitive DNA PCR A clone fingerprinting technique that uses PCR to detect the relative positions of genome-wide repeats in cloned DNA fragments.

repetitive elements Repetitive elements provide important clues about chromosome dynamics, evolutionary forces, and mechanisms for exchange of genetic information between organisms. The most ubiquitous class of repetitive elements in the DNA sequence in primate genomes is the *Alu* family of interspersed repeats which have arisen in the last 65 million years of evolution. *Alu* repeats belong to a class of sequences defined as short interspersed elements (SINEs). Approximately 500,000 *Alu* SINEs exist within the human genome, representing about 5% of the genome by mass.

replacement The result of a non-synonymous substitution at the protein level.

replica plating A technique for transfer of colonies from one Petri dish to another, such that their relative positions on the surface of the agar medium are retained.

replication *See* DNA replication.

Replication Factor C (RFC) A multi-subunit accessory protein involved in eukaryotic DNA replication.

replication factory A large structure attached to the nuclear matrix; the site of DNA replication.

replication fork The region of a double-stranded DNA molecule that is being opened up to enable DNA replication to occur.

Replication Licensing Factors (RLFs) A set of proteins that regulate DNA replication, by ensuring that only one round of DNA replication occurs per cell cycle.

replication origin A site on a DNA molecule where replication initiates.

Replication Protein A (RPA) The main single-strand binding protein involved in replication of eukaryotic DNA.

replication slippage An error in replication that leads to an increase or decrease in the number of repeat units in a tandem repeat such as a microsatellite.

replicative transposition Transposition that results in copying of the transposable element.

replicator gene A regulatory gene specifying the sites for initiation and termination of DNA replication.

replicon A chromosomal region that contains the DNA sequences necessary for the initiation of DNA replication, and that is replicated as a unit.

replisome A complex of proteins involved in DNA replication.

reporter gene A gene whose phenotype can be assayed to determine the function of a regulatory DNA sequence.

repository A collection of resources that can be accessed to retrieve information.

repressor The protein product of a regulatory gene that combines with a specific operator (regulatory DNA sequence) and hence blocks the transcription of genes in an operon.

representation The method chosen to model a process or object. For example, a building may be represented as a physical scale model, drawing or photograph.

reproductive barrier Any of several biological or environmental mechanisms that prevent gene exchange between population.

Research Collaboratory for Structural Bioinformatics (RCSB) A collaborative structural biology research group that maintains the Protein Data Bank (PDB).

Research in Computational Molecular Biology (RECOMB) A well established conference in a dynamic field of science bridging computer science and biology.

residue A single unit in a polymer. It is used for both a single nucleotide in DNA or a single amino acid in a protein.

resolution Degree of molecular detail on a physical map of DNA, ranging from low to high.

restriction enzyme An enzyme that recognizes specific, short nucleotide sequences and cuts DNA at those sites.

restriction enzyme cutting site A specific nucleotide sequence of DNA at which a particular restriction enzyme cuts the DNA.

restriction fragment A (relatively) short fragment of DNA which is the result of the digestion of high-molecular weight DNA by a restriction endonuclease.

Restriction Fragment Length Polymorphism (RFLP) A heritable difference in DNA fragment length and fragment number; passed from generation to generation in a co-dominant way.

restriction map A physical map or depiction of a gene (or genome) derived by ordering the overlapping restriction fragments produced by digestion of the DNA with a number of restriction enzymes.

restriction mapping Determination of the positions of restriction sites in a DNA molecule by analysing the sizes of restriction fragments.

restriction site The location on a DNA or protein chain at which a specific restriction enzyme will act.

restrictive conditions Conditions under which a conditional-lethal mutant is unable to survive.

retroelement A genetic element that transposes via an RNA intermediate.

retrofection Transfer of an RNA molecule from one cell to another, in particular to a germline cell, by means of a retroviral particle, into which the RNA is encapsulated.

retrogene A functional retrosequence producing a protein that is identical or nearly identical to that produced by the gene from which an mRNA was derived.

retron A genomic sequence encoding reverse transcriptase but lacking the ability to transpose.

retroposon A transposable retroelement that neither constructs virion particles nor is flanked by terminally redundant sequences.

retrosequence A genomic sequence that has been derived through the reverse transcription of RNA but by itself lacks the ability to produce reverse transcriptase.

retrotransposition The integration of a sequence derived from an RNA into a DNA genome.

retrotransposon A genome-wide repeat with a sequence similar to an integrated retroviral genome and possibly with retrotransposition activity.

retroviral infection The presence of retroviral vectors, such as some viruses, which use their recombinant DNA to insert their genetic material into the chromosomes of the host's cells.

Retroviral-like element (RTVL) A truncated retroviral genome integrated into a host chromosome.

retrovirus A virus with an RNA genome that integrates into the genome of its host cell.

reverse genetics An old name for positional cloning.

reverse transcriptase A polymerase that synthesizes DNA on an RNA template.

reverse transcription Process of transcribing a single-stranded DNA from a single-stranded RNA (the reverse of transcription); used by retroviruses as well as in biotechnology.

Reverse Transcription PCR (RT-PCR) PCR in which the first step is carried out by reverse transcriptase, so RNA can be used as starting material.

RH mapping A statistical method used to determine the distance between DNA markers, as well as their order on the chromosome. This technique uses X-rays to break the chromosome.

Rho-dependent terminator A position in bacterial DNA where termination of transcription occurs with the involvement of Rho.

Rho helicase An enzyme required for termination of some bacterial transcripts.

ribbon-helix-ribbon motif A type of DNA-binding domain.

ribonuclease An enzyme that catalyses (breaks down) RNA.

ribonuclease D A ribonuclease enzyme involved in processing pre-tRNA in bacteria.

ribonuclease P An enzyme that is involved in processing pre-tRNA in bacteria.

ribonucleic acid (RNA) Nucleotide made from a ribose, a base [adenine (A), guanine (G), cytosine (C), and uracil (U)], and a phosphate group. RNA is generally found in the cell nucleus or cytoplasm. The three types of RNA are messenger RNA (mRNA), transfer RNA (tRNA) and ribosomal RNA (rRNA).

ribonucleoprotein domain A common type of RNA-binding domain.

ribose The five-carbon sugar found in nucleotides of RNA and in ATP.

ribosomal protein One of the protein components of a ribosome.

ribosomal RNA (rRNA) A class of RNA found in the ribosomes of cells.

ribosome Cellular components made of ribosomal RNA and proteins which are the site of protein synthesis (translation).

ribosome binding site The nucleotide sequence that acts as the attachment site for the small subunit of the ribosome during initiation of translation in bacteria.

ribozyme An RNA molecule that has catalytic activity.

risk communication In genetics, a process in which a genetic counsellor or other medical professional interprets genetic test results and advices patients of the consequences for them and their offspring.

RNA-dependent DNA polymerase An enzyme that makes a DNA copy of an RNA template; a reverse transcriptase.

RNA-dependent RNA polymerase An enzyme that makes an RNA copy of an RNA template.

RNA editing A process by which nucleotides, not coded by a gene, are introduced at specific positions in an RNA molecule after transcription.

RNA polymerase An enzyme that synthesizes RNA on a DNA or RNA template.

RNA polymerase I The eukaryotic RNA polymerase that transcribes ribosomal RNA genes.

RNA polymerase II The eukaryotic RNA polymerase that transcribes protein-coding and snRNA genes.

RNA polymerase III The eukaryotic RNA polymerase that transcribes transfer RNA and other short genes.

RNA transcript An RNA copy of a gene.

RNA world The hypothesized early period of evolution when all biological reactions were centred on RNA.

rolling circle replication A replication process that involves continual synthesis of a polynucleotide which is "rolled-off" of a circular template molecule.

root In rooted trees, the common ancestor of all the taxa under study.

rooted A phylogenetic tree that provides information on the past evolutionary events that have led to the organisms or DNA sequences being studied.

rooted tree A phylogenetic tree that specifies ancestral and descendant species, thus indicating the direction of the evolutionary path.

Root Mean Square Deviation (RMSD) A method in molecular modelling of measuring the difference between different configurations of the same molecule.

row In a relational database, one set of attributes (or one tuple) corresponding to one instance of the entity that a table schema describes.

S

saddle point A point on a surface that is neither a peak nor a valley, but still has a 0 gradient.

Sanger sequencing A widely used method of determining the order of bases in DNA.

satellite A chromosomal segment that branches off from the rest of the chromosome but is still connected by a thin filament or stalk.

satellite DNA Repetitive DNA that forms a satellite band in a density gradient.

Scaffold Attachment Region (SAR) An AT-rich segment of a eukaryotic genome that acts as an attachment point to the nuclear matrix.

scalar A single number, as opposed to a multidimensional vector or matrix.

scanning The process during initiation of eukaryotic translation in which the preinitiation complex attaches to the 5´-terminal cap structure of an mRNA and then scans along the molecule until it reaches an initiation codon.

schema A similarity template used to analyse genetic algorithms. By using wild-card characters, a schema defines an entire class of strings that may be found in a population.

score matrix In a dynamic programming alignment, the score matrix indicates the quality of an alignment ending at each possible pair of residues.

scoring function The methods used to evaluate the quality of the overlap between sequences. A variety of scoring functions are used to evaluate single replacement operations, multiple alignments (either whole or columns), and pairwise alignments. The score of an alignment of two sequences (a and b) is the sum of the score of all the replacement operations that lead from a to b. It is also called as cost function or weight function.

search A method for finding a region of interest in a search space.

search engine Computer program capable of seeking information on the World Wide Web (or indeed any large database) based upon search criteria specified by a user.

search space The set of possible solutions to a problem instance.

secondary extinction The death of one population due to the extinction of another, often a food species.

secondary structure The organization of the peptide backbone of a protein that occurs as a result of hydrogen bonds, e.g. alpha-helix, beta-pleated sheet.

second law of thermodynamics The energy available after a chemical reaction is less than that at the beginning of a reaction; energy conversions are not 100% efficient.

second messenger An intermediate in a certain type of signal transduction pathway.

Second Normal Form (2NF) A relation that is in first normal form and where each non-key attribute in the relation is functionally dependent upon the primary key.

seed alignment An alignment that contains only one of each pair of homologues that are represented in a ClustalW-derived phylogenetic tree linked by a branch of length less than a distance of 0.2.

SEG A program for filtering low- complexity regions in amino acid sequences. Residues that have been masked are represented as "X" in an alignment. SEG filtering is performed by default in the Blastp subroutine of BLAST 2.0.

segment polarity genes Developmental genes that provide greater definitionnmmmmmmm- mmmmm to the segmentation pattern of the *Drosophila* embryo established by the action of the pair-rule genes.

segregation The separation of homologous chromosomes, or members of allele pairs, into different gametes during meiosis.

segregator gene A regulatory gene providing chromosomal

attachment sites for the segregation machinery during meiosis and mitosis.

selection coefficient A quantitative measure of the reduction in fitness of a genotype in comparison with the fittest genotype in the population.

selection intensity The difference in the fitness values between the various genotypes in a population.

selective breeding The selection of individuals with desirable traits for use in breeding. Over many generations, the practice leads to the development of strains with the desired characteristics.

selective medium A medium that supports the growth of only those cells that carry a particular genetic marker.

selectivity Selectivity of bioinformatics similarity search algorithms is defined as the significance threshold for reporting database sequence matches. As an example, for BLAST searches, the parameter E is interpreted as the upper bound on the expected frequency of chance occurrence of a match within the context of the entire database search. The parameter E may be thought as the number of matches one expects to observe by chance alone during the database search.

selfish DNA DNA that appears to have no function and apparently contributes nothing to the cell in which it is found.

selfish genetic element A very diverse and common feature of eukaryotic genomes, such as transposons.

self-organization A spontaneously formed higher-level pattern of structure or function that is emergent through the interactions of lower level objects.

Self-Organized Criticality (SOC) A mathematical theory that describes how the systems composed of many interacting parts can tune themselves toward dynamical behaviour that is critical in the sense that it is neither stable nor unstable.

self-similar An object whose parts resemble the whole.

self-splicing intron An intron that can be cleaved out of the pre-mRNA without the aid of an internal catalyst.

semantics The meaning associated with a set of symbols in a given language, which is determined by the syntactic structure of the symbols, as well as knowledge captured in an interpretative model.

semiconservative replication The mode of DNA replication in which each new double helix is made up of one polynucleotide from the parent and the other one is newly synthesized polynucleotide.

sense codon A codon specifying an amino acid.

sense strand The non-transcribed strand of a gene, the DNA sequence which is identical to the RNA transcript.

sensitivity The probability that a test will be positive, given that the condition to be detected is present. Sensitivity of bioinformatics similarity search algorithms centres around two areas: First, how well can the method detect biologically meaningful relationships between two related sequences in the presence of mutations and sequencing errors; Secondly how does the heuristic nature of the algorithm affect the probability that a matching sequence will not be detected. At the user's discretion, the speed of most similarity search programs can be sacrificed in exchange for greater sensitivity— with an emphasis on detecting lower scoring matches.

sensor In a closed system, an element that detects change and signals the effector to initiate a response.

sequence assembly A process whereby the order of multiple sequenced DNA fragments are determined.

sequence divergence The differences between two homologous sequences due to the independent accumulation of genetic changes in each lineage.

sequencer An apparatus used for deciphering the order of bases in a DNA strand.

Sequence Retrieval System (SRS) Program for biological database, created by T. Etzold (EMBL).

sequence skimming A method for rapid sequence acquisition in which a few random sequences are obtained from a cloned fragment, the rationale being that if the fragment contains any genes then there is a good chance that at least some of them will be revealed by these random sequences.

Sequence Tagged Site (STS) A short (200 to 500-bp) DNA sequence that has a single, unique occurrence in an organism's genome, and whose location and base sequence are known.

sequencing Determination of the order of nucleotides (base sequences) in a DNA or RNA molecule or the order of amino acids in a protein.

sequencing technology The instrumentation and procedures used to determine the order of nucleotides in DNA.

sequin A stand-alone software tool developed by NCBI for submitting and updating entries to the GenBank, EMBL, or DDBJ sequence databases.

serine (Ser, S) One of the 20 naturally occuring proteinogenic amino acids. Its codons are UCU, UCC, UCA, UCG, AGU and AGC.

server A computer that processes requests issued from remote locations by client machines.

set A collection of objects, in which each object appears once.

sex chromosome A chromosome involved in sex determination; in humans, the X and Y chromosomes are the sex chromosomes.

sex-influenced A trait from the same alleles which is expressed to different degrees depending on the sex of the individual.

sex-limited A trait which is not usually sex-linked but is only expressed in one sex.

sex linkage The situation in which a gene is located on a sex chromosome.

sex-linked A trait which is caused by mutation of a gene on the X chromosome, usually called X linkage to avoid confusion with Y-linked traits.

sexual PCR Sexual PCR is a form of PCR in which similar, but not identical, DNA sequences are reassembled to obtain novel juxtapositions, simulating the result of genetic recombination. The result is the creation of an array of related genes which may possess improved characteristics. By repeated rounds of recombination, selection and PCR-based amplification vastly improved gene-products, such as enzymes with greater activity, may be generated and selected.

shadowing lemma A numerical simulation of chaos may "shadow" a real trajectory of a real chaotic system.

Shine–Dalgarno Sequence The ribosome binding site, i.e., upstream of a gene in *E. coli.*

Short Interspersed Nuclear Element (SINE) A type of genome-wide repeat, less than 500 bp in length.

short patch repair A repair process of *E. coli* that results in excision and resynthesis of about 12 nucleotides of DNA.

shotgun approach A genome-sequencing strategy in which the molecules to be sequenced are randomly broken into fragments which are then individually sequenced.

shotgun cloning The cloning of an entire gene segment or genome by generating a random set of fragments using restriction endonucleases to create a gene library that can be subsequently mapped and sequenced to reconstruct the entire genome.

shotgun method A method that usesenzymestocutDNAintohundreds (or thousands) of random bits which are then reassembled by computer so it looks like the original genome. The Human Genome Project (HGP) shotgun approach is applied to cloned DNA fragments that already have been mapped so that it is known exactly where they are located on the genome, making assembly easier and much less prone to error.

shotgun sequencing An approach used to decode an organism's genome by shredding it into smaller fragments of DNA which can be sequenced individually.

sibling A brother or sister.

sibling species Species that are indistinguishable morphologically but are reproductively isolated.

sibship All the brothers and sisters in one family.

sickle-cell anaemia Human autosomal recessive disease that causes production of abnormal red blood cells that collapse (or sickle) and cause circulatory problems.

sigmoidal function An "S"-shaped function that is often used as an activation function in a neural network.

signal peptide A short sequence at the N-terminus of some proteins that directs the protein across a membrane.

signal sequence (leader sequence) A short sequence added to the amino-terminal end of a polypeptide chain that forms an amphipathic helix allowing the nascent polypeptide to migrate through membranes such as the endoplasmic reticulum or the cell membrane. It is cleaved from the polypeptide after the protein has crossed the membrane.

Signal Transducer and Activator of Transcription (STAT) A type of protein that activates a transcription factor in response to binding of an extracellular signalling compound to a cell surface receptor.

signal transduction A biological process that occurs when an extracellular signal causes an intracellular signal resulting in a physiological response.

silencer A regulatory sequence that reduces the rate of

transcription of a gene or genes located some distance away in either direction.

silent allele An allele that has no effect on the phenotype.

silent mutation A change in a DNA sequence that has no effect on the expression or functioning of any gene or gene product.

similarity The extent to which nucleotide or protein sequences are related. The extent of similarity between two sequences can be based on per cent sequence identity and/or conservation. In BLAST similarity refers to a positive matrix score.

similarity approach A rigorous mathematical approach to alignment of nucleotide sequences.

similarity (homology) search Given a newly sequenced gene, there are two main approaches to the prediction of structure and function from the amino acid sequence. Homology methods are the most powerful and are based on the detection of significant extended sequence similarity to a protein of known structure, or of a sequence pattern characteristic of a protein family. Statistical methods are less successful but more general and are based on the derivation of structural preference values for single residues, pairs of residues, short oligopeptides or short sequence patterns. The transfer of structure/ function information to a potentially homologous protein is straightforward when the sequence similarity is high and extended in length, but the assessment of the structural significance of sequence similarity can be difficult when sequence similarity is weak or restricted to a short region.

similarity measure (similarity function or similarity score) A function that associates a numeric value with (a pair of) sequences, with the idea that a higher value indicates greater similarity. A similarity measure is a type of scoring function that is used to rank the degree of similarity of a pair of sequences.

simple graph A graph that contains no loops (i.e., edges of the form (v, v) for any v in V) or multiple edges between the same pair of vertices.

Simple Sequence Length Polymorphism (SSLP) An array of repeat sequences that display length variations.

simulated annealing A method for finding solutions to combinatorial optimization problems.

single-copy DNA A DNA sequence that is not repeated elsewhere in the genome.

single-gene disorder A hereditary disorder caused by a mutant allele of a single gene, such as Duchenne muscular dystrophy, retinoblastoma, sickle-cell disease.

Single Nucleotide Polymorphism (SNP) Variations of single base pairs scattered throughout the human genome that serve as measures of the genetic diversity in humans. About 1 million SNPs are estimated to be present in the human genome, and SNPs are useful markers for gene mapping studies.

single orphan A single gene, with no homolog, whose function is unknown.

single-pass sequencing Rapid sequencing of large segments of the genome of an organism by isolating as many expressed (cDNA) sequences as possible and performing single sequencer runs on their 5′ or 3′ ends. Single-pass sequencing typically results in individual, error-prone sequencing reads of 400–700 bases, depending on the type of sequencer used. However, if many of these are generated from numerous clones from different tissues, they may be overlapped and assembled to remove the errors and generate a contiguous sequence for the entire expressed gene.

Single-Strand Binding Protein (SSB) One of the proteins that attach to single-stranded DNA in the region of the replication fork, preventing base pairs forming between the two parent strands before they have been copied.

single-stranded A DNA or RNA molecule that consists of a single polynucleotide chain.

sister chromatid exchange A crossover between sister chromatids (chromatids which are the products of replication of a single chromosome and which ought therefore to be genetically identical).

sister chromatids Chromatids joined by a common centromere and carrying identical genetic information (unless crossing-over has occurred).

sister taxa The pair of species among a group of species, that are evolutionarily close to each other.

site An individual column of residues in an amino acid or nucleotide alignment. The residues at a site are presumed to be homologous.

site-directed hydroxyl radical probing A technique for locating the position of a protein in a protein-RNA complex, such as a ribosome, by making use of the ability of Fe(II) ions to generate hydroxyl radicals which cleave nearby RNA phosphodiester bonds.

site-directed mutagenesis Techniques used to produce a specified mutation at a pre-determined position in a DNA molecule.

site-specific recombination Recombination between two double-stranded DNA molecules that have only short regions of sequence similarity.

slippage The translocation of a ribosome along a short non-coding nucleotide sequence between the stop codon of one gene and the start codon of a second gene.

SMAD family A group of proteins involved in signal transduction.

small cytoplasmic RNA (scRNA) A type of short RNA molecule with various roles in eukaryotic cell.

small nuclear Ribonucleo-protein (snRNP) A compound of one or two snRNA molecules complexed with proteins.

small nuclear RNA (snRNA) A type of short RNA molecule, found in eukaryotes, involved in splicing GU–AG and AU–AC introns and other types of RNA processing.

small nucleolar RNA (snoR-NA) A type of short eukaryotic RNA molecule involved in chemical modification of rRNA.

SNOMED A commercially available general medical terminology, initially developed for the classification of pathological specimens.

Sl nuclease An enzyme that degrades single-stranded DNA or RNA molecules including single-stranded regions in predominantly double-stranded molecules.

solvent accessibility The surface area (typically measured in square angstroms) of a biological molecule, usually a protein, that is exposed to solvent in its native, folded form. Determining the solvent accessibility of a protein helps to define which amino acids in its molecular sequence are on the exterior of the molecule, and thus available to participate in interactions with other molecules.

somatic cell Any cell in the body except gametes (sex cells, such as ova and spermatozoa) and their precursors.

somatic cell gene therapy Incorporating new genetic material into cells for therapeutic purposes. The new genetic material cannot be passed to offspring.

somatic cell genetics The study of genes using hybrids (fusions) between the somatic cells of different species.

somatic mutation   A mutation occurring in a somatic cell, which is neither inherited nor passed to offspring. That is mutations that occur during *in utero* development.

SOS response   A series of biochemical changes that occur in *E. coli* in response to damage to the genome.

Southern blot   A technique used to identify and locate DNA sequences which are complementary to another piece of DNA called a probe.

Southern hybridization   A technique used for detection of a specific restriction fragment against a background of many other restriction fragments.

space complexity   A function that describes the amount of memory required for a program to run on a computer to perform a particular task.

space-filling curve   A curve that twists and turns in such a way that it fills a space or volume or two or more dimensions.

spacer DNA   The DNA found between two genes; can be either transcribed or non-transcribed.

spam   Postings to newsgroups or mail broadcast to a large number of e-mail accounts that usually are irrelevant or not of interest to the recipients. Analogous to postal junk mail.

spanning tree   A connected, acyclic subgraph containing all the vertices of a graph.

spatio-temporal dynamics   Local interactions in space can give rise to large-scale spatio-temporal patterns (e.g. (spiral) waves, spatio-temporal chaos (turbulence), stationary (Turing-type) patterns and transitions between these modes). Their occurrence and properties are largely independent of the precise interaction structure. They are indeed seen to occur at many organizational levels of biotic systems. Space can be either "real" space or a state space, e.g. "phenotype space" in models of speciation or "shape space" in immunological models of shape- based receptor interactions.

speciation   The splitting of one population into two or more populations that are reproductively isolated; the process by which new species arise.

species A basic taxonomic category.

species diversity The number of living species on earth.

species richness   The number of species present in a community.

species tree A phylogenetic tree that shows the evolutionary relationships between a group of species.

specificity The probability that a test will correctly detect the absence of an underlying condition.

Spectral Karotype (SKY) A visualization of all of an organism's chromosomes together, each labelled with a different colour. This technique is useful for identifying chromosome abnormalities.

sperm The male gamete.

S phase A stage of the cell cycle when DNA synthesis occurs.

spindle The structure of microtubules that pulls the chromosomes into the daughter cells at anaphase in cell division.

splice form By using alternative splicing, a single message precursor from DNA can generate an entire family of mRNAs and proteins. This can be utilized to create specificity in cell–cell or cell–ligand interactions. A cell may produce a given protein, but it will be a different splice-form of the protein than that produced by an adjacent cell. In this manner, the two cells have the potential to interact differently with other cells or molecules.

splice junction mutation A mutation that alters a junction between an intron and an exon so that it no longer functions properly.

spliceosome The protein–RNA complex involved in splicing GU–AG or AU–AC introns.

splice site Location in the DNA sequence where RNA removes the non-coding areas to form a continuous gene transcript for translation into a protein.

splicing The process of cutting, excising, and recombining an RNA or DNA. In RNA, splicing is used to remove introns from the coding sequence.

splicing enhancer A nucleotide sequence that plays a regulatory role during splicing of GU–AG introns.

spontaneous mutation A mutation that arises from an error in replication.

sporadic A case of a genetic disease caused by a new mutation.

sporadic cancer Cancer that occurs randomly and is not inherited from parents.

SR protein A protein that plays a role in splice-site selection during splicing of GU–AG introns.

SRSWWW Version of the Sequence Retrieval System (SRS) available through the World Wide Web.

stabilizing selection A process of natural selection that tends to favour genotypic combinations that produce an intermediate phenotype; selection against the extremes in variation.

stable A basin of attraction that is non-zero in size; an attractor that can withstand some form of perturbation.

Standard Deviation (SD) A measure of the spread or dispersion of a set of values; for variables that follow a normal distribution, the probability that a value will be within one standard deviation of the mean is about 2/3.

Standard Generalized Markup Language (SGML) An international standard meta-language for defining document definition languages. SGML forms the basis for HTML, XML, and other markup languages.

start codon The codon, usually but not exclusively AUG, that signals the start of a gene sequence to be translated to protein.

state The condition of a finite state machine or Turing machine at a certain time.

statement In a formal system, a string of characters that is formed according to well-defined rules such that it is legal for the language that is the formal system.

state space A set of parameters that describe the behaviour of a dynamical system.

steady state A zero-dimensional (point) attractor.

steepest descent A search method that uses the gradient information of a search space and moves in the opposite direction from the gradient until no further downhill progress can be made.

stem cell An undifferentiated, primitive cell in the bone marrow that has both the ability to multiply and to differentiate into specific blood cells.

stem-loop structure A structure made up of a base-paired stem and nonbase-paired loop, which can form in a single-stranded polynucleotide that contains an inverted repeat.

steric hindrance Atoms in a molecule are forced into certain configurations because other parts of the molecules are occupying nearby regions in space and physically blocking certain atoms from exploring those regions.

steroid hormone A type of extracellular signalling compound.

steroid receptor A protein that binds a steroid hormone after the latter has entered the cell, as an intermediate step in modulation of genome activity.

sticky end An end of a double-stranded DNA molecule where there is a single-stranded extension.

stochastic process A process, the outcome of which cannot be predicted exactly from knowledge of initial conditions.

stop codon One of the three codons that mark the position where translation of an mRNA should stop.

strange attractor An attractor of a dynamical system that is usually fractal in dimension and is indicative of chaos.

strategy In game theory, a policy for playing a game.

strength For a classifier system, a classifier's relative ability to win a bidding match for the right to post its message on the message list.

stress-activated protein kinase A signal transduction pathway activated by cellular stress.

string An ordered succession of characters or symbols drawn from a finite alphabet.

stringency When a statistic or score is calculated for a window of residues, the term stringency is used to refer to the minimum score which will be saved or considered to be a match.

strong bond In reference to double-stranded nucleic acids or segments, the three hydrogen bonds between C and G confers increased stability and high melting temperature.

strong promoter A promoter that directs a relatively large number of productive initiations per unit time.

structural domain A segment of a polypeptide that folds independently of other segments.

structural gene A DNA nucleotide sequence that codes for a protein or specifies an RNA molecule. Gene which encodes a structural protein.

structural genomics The prediction of the 3D structure of proteins encoded by genes using both experimental and computational techniques.

structure-based drug design Pharmaceutical research

driven by the three-dimensional structure of a target site.

structured data The complex and richly structured data from genomics can be viewed as the greatest encoding problem of all time. From this perspective, the sequencing of the human and other genomes can be viewed as one of the all-time great opportunities for theorists interested in information, its structure and analysis.

Structured Query Language (SQL) An industry-standard language for creating, updating and querying relational database management systems.

structure prediction Using algorithms to predict the secondary, tertiary, and quaternary structure of proteins based on the genomic sequences from which they are formed.

STS mapping A physical mapping procedure that locates the positions of sequence-tagged sites (STSs) in a genome.

stuffer fragment A DNA fragment contained within a lambda vector that is replaced by the DNA to be cloned.

subgraph A graph composed of a subset of the parent graph's nodes and edges.

subsequence A string formed by removing zero or more characters from the given string.

subspecies A geographically or morphologically distinct population in a species.

substitution The presence of a non-identical amino acid at a given position in an alignment. If the aligned residues have similar physico-chemical properties the substitution is said to be "conservative".

substitution matrix A matrix containing values proportional to the probability that amino acid i mutates into amino acid j for all pairs of amino acids. Such matrices are constructed by assembling a large and diverse sample of verified pairwise alignments of amino acids. If the sample is large enough to be statistically significant, the resulting matrices should reflect the true probabilities of mutations occurring through a period of evolution.

substring A subsequence of a string formed by consecutive characters in the original string.

subtraction library A cDNA library that only contains cDNAs uniquely expressed in a given cell or tissue, e.g. T cells and B cells will express many common RNAs, as well as a very small percentage which will be unique for T cells and B cells respectively. To make a T-cell subtraction library, the cDNA from a T-cell library is hybridized with a vast excess of B-cell RNA. The commonly expressed genes will result in RNA–cDNA hybrids which can be removed (or subtracted) to leave only T-cell specific cDNAs.

suffix A substring of characters at the end of a string.

suicide gene A strategy for making cancer cells more vulnerable to chemotherapy.

supercoiling A conformational state in which a double helix is overwound or underwound so that superhelical coiling occurs.

superfamily A collection of genes, all products of gene duplication, that have diverged from each other to a considerable extent (in protein-coding genes, usually a similarity of less than 50% at the amino acid level).

superkey A combination of attributes that can be uniquely used to identify a database record.

suppressor gene A gene that can suppress the action of another gene.

suppressor mutation A mutation in one gene that reverses the effect of ("suppresses") a mutation in a second gene.

survival State of surviving. Remaining alive; a natural process resulting in the evolution of organisms best adapted to the environment.

S-value The unit of measurement for a sedimentation coefficient.

Swiss Institute of Bioinformatics (SIB, ISB) An academic institution established on March 30, 1998 as a non-profit foundation. The goals of this institute are to promote the development of software tools and databases in the field of bioinformatics, to sustain a high-quality research program in bioinformatics, to provide, in collaboration with academic partners, a curriculum of courses and seminars for the formation of research scientists in the field of bioinformatics, and to offer services to the Swiss scientific user community through the Swiss-EMBnet node (which is currently maintained jointly by Swiss Institute for Cancer Research and the University of Geneva).

The Swiss Institute of Bioinformatics operates three web servers: the ExPASy proteomics server, the Swiss node of EMBnet, and "Protéine à la Une".

Swiss-Prot A non-redundant protein sequence database. An annotated protein sequence database established in 1986 and maintained collaboratively, since 1987, by the Department of Medical Biochemistry of the University of Geneva and the EMBL Data Library (now the EMBL Outstation—The European Bioinformatics Institute (EBI)). The Swiss-Prot protein sequence data bank consists of sequence entries. Sequence entries are composed of different line-types, each with their own format. For standardization purposes the format of Swiss-Prot follows as closely as possible that of the EMBL nucleotide sequence database.

symbiosis An interactive relationship between two or more species living together; may be parasitic, commensal, or mutualistic.

symmetrical exon An exon residing between two same-phase introns.

symmetric matrix A matrix whose lower-left half is equal to the mirror image of its upper-right half.

synapse The junction between two neurons in which neural activity is propagated from one neuron to another.

synapsis The close pairing of homologous chromosomes in meiotic prophase which enables crossovers to occur.

synaptonemal complex The electron dense material which acts as a "zip fastener" to hold together homologous chromosomes in meiotic prophase.

synchronous Acting in a lockstep fashion, with each event occurring in a precise order, or in such a way as to eliminate the notion of order entirely.

synchronous communication A mode of communication when two parties exchange messages across a communication channel at the same time, e.g. by telephone.

syncytium A cell-like structure comprising a mass of cytoplasm and many nuclei.

syndrome The group or recognizable pattern of symptoms or abnormalities that indicate a particular trait or disease.

syngeneic Genetically identical members of the same species.

synonymous mutation A mutation that changes a codon into a second codon that specifies the same amino acid.

syntax The rules of grammar that define the formal structure of a language.

synteny A pair of genomes in which at least some of the genes are located at similar map positions.

synteny map Map indicating whether or not genes occur in the same order on chromosomes of different species.

system Something that can be studied as a whole.

systematics The classification of organisms based on information from observations and experiments; includes the reconstruction of evolutionary relatedness among living organisms.

systems bioinformatics With the completion of the human genome project, the scientific community is now faced with the even greater challenge of analysing the resulting data from this and other large-scale genome projects to better understand the networks underlying biological function.

Systems Biology Markup Language (SBML) A model-based description language for systems biology simulation software.

T

table A predefined format of rows and columns in a relational database that defines an entity.

tandem repeat Direct repeats that are adjacent to each other.

tandem repeat sequences Multiple copies of the same base sequence on a chromosome. These sequences are used as markers in physical mapping.

tar A UNIX command for copying files to or restoring files from a backup tape or other storage medium.

tarball A collection of files compressed into one using the "tar" command.

target A molecule that is critical to a disease that may be targeted with a potential therapeutic agent.

target gene A gene of part of the gene that is being studied.

TATA-Binding Protein (TBP) A component of the general transcription factor TFllD, the part that recognizes the TATA box of the RNA polymerase II promoter.

TATA box Part of the RNA polymerase II core promoter.

tautomer One of the two or more structural isomers in dynamic equilibrium.

tautomeric shift The spontaneous change of a molecule from one structural isomer to another.

taxon A taxonomic group of any rank (e.g. species, genus, kingdom) to which individual organisms are assigned.

taxonomy The principles and procedures according to which species are named and assigned to taxonomic groups.

TBP-Associated Factor (TAF) One of the several components of the general transcription factor TFIID, playing ancillary roles in recognition of the TATA box.

TBP domain A type of DNA-binding domain.

TΨC arm Part of the structure of a tRNA molecule.

telomere The free end of a chromosome.

Tentative Consensus (TC) The identification of a sequence from an EST cluster that represents part or all of a complete gene. TCs are usually determined by clustering ESTs allowing for sequencing errors, artefacts such as chimeric clones, and naturally occuring biological phenomena such as alternative splicing. Creation of a cluster allows one to generate a consensus sequence and then identify a long open reading frame which would suggest the possibility of that consensus representing a bona fide gene.

Tentative Human Consensus sequences (THCs) A consensus sequence generated from human EST fragments. THCs may be validated by comparison against databases of known human gene sequences, human genomic sequences, or by identification of the ORFs or other sequence features contained within the consensus as belonging to a known human gene product.

tertiary structure Folding of a protein chain via interactions of its side chain molecules including formation of disulphide bonds between cysteine residues.

threading Computational algorithms which use the known 3D structure of proteins as a template for positioning (or "threading") an unknown sequence. The overlap of the sequences is based on how the sequences match with respect to physical properties at specific residues rather than matching based on the similarity of amino acid sequences.

thymine A pyrimidine base found in DNA but not in RNA.

TIGR The Institute for Genomic Research was a non-profit genomics research institute founded in 1992 by Craig Venter in Rockville, Maryland, United states. It is now a part of the J.Craig Venter Institute.

tissue Section of an organ that consists of a largely homogenous population of cell types. Since many organs are multifunctional, they have developed highly specialized cell types to perform different functions. Identifying the section of an organ that is homogenous for a particular cell type ensures that the gene expression profiles extracted from those cells will accurately resemble the class of cells that make up the tissue.

trace A series of coloured peaks from which the individual bases of a sequence are derived. The original format is produced by the ABI analysis software. This format is converted to SCF for use by Xgap, by the Squirrel program.

transcript The single-stranded mRNA chain that is assembled from a gene template.

transcription The process of copying a strand of DNA to yield a complementary strand of RNA.

transcription factors A group of regulatory proteins that are required for transcription in eukaryotes. Transcription factors bind to the promoter region of a gene and facilitate transcription by RNA polymerase.

transcriptome The complete collection of RNA molecules transcribed (or processed) from the DNA of a cell.

transfection Introduction of a foreign DNA molecule into a eukaryotic cell and subsequent expression of the genes of the new DNA.

transfer RNA (tRNA) A small RNA molecule that recognizes a specific amino acid, transports it to a specific codon in an mRNA, and positions it properly in the nascent polypeptide chain.

transformation A genetic alteration to a cell as a result of the incorporation of DNA from a genetically different cell or virus. It can also refer to the introduction of DNA into bacterial cells for genetic manipulation.

transgene A foreign gene that is introduced into a cell or whole organism (e.g. transgenic mice) for therapeutic or experimental purposes.

translation The process of converting RNA to protein by the assembly of a polypeptide chain from an mRNA molecule at the ribosome.

transmembrane region The region of a transmembrane protein that actually spans the membrane, i.e., trans-membrane regions are usually hydrophobic in order to be thermo-dynamically compatible with the lipid bilayer portion of the membrane. They may consist of either alpha-helical or beta-strand secondary structure elements, but in either case the external residues (the ones facing the membrane) are invariably hydrophobic while the internal residues may be hydrophilic (as in the case of a pore or channel) or polar. One common transmembrane structural domain is the seven-helix bundle seen in numerous channel proteins.

TrEMBL   The supplement of Swiss-Prot that contains all the translations of EMBL nucleotide sequence entries not yet integrated in Swiss-Prot.

U

Ultracentrifuge A centrifuge capable of speeds of rotation sufficiently high to separate molecules of different density in a suitable density gradient.

Uncharged tRNA A tRNA molecule that lacks an amino acid.

Uncovering The expression of recessive allele present in a region of a structurally normal chromosome in which the homologous chromosome has a deletion.

Underwound DNA A DNA molecule whose strands are untwisted somewhat; hence, some of its bases are unpaired.

Unequal crossing over Crossing over between non-allelic copies of duplicated or other repetitive sequences in a tandem duplication, between the upstream copy in one chromosome and the downstream copy in the homologous chromosome.

Unidentified Reading Frame (URF) An open reading frame encoding a protein of undefined function.

Uniparental Inheritance Extranuclear inheritance of a trait through cytoplasmic factors or organelles contributed by only one parent.

Unique sequence A DNA sequence that is present in only one copy per haploid genome, in contrast with repetitive sequences.

unitary matrix (Identity Matrix) A scoring system in which only identical characters receive a positive score.

Univalent Structure formed in meiosis I in a monoploid or a monosomic when a chromosome has no pairing partner.

Unlinked genes Genes whose frequency of recombination is 50 per cent (independent assortment); unlinked genes can be in the same chromosome if they are sufficiently far apart.

Untranslated Region (UTR) A portion of the mRNA that is not utilized, to produce protein and is located upstream of the start codon (5′ UTR) and downstream of the stop codon (3′ UTR).

Upstream   A relative direction for nucleic acids often used to describe the location of a promoter relative to the start transcription site, e.g. the start codon is upstream of the stop codon.

uracil Nitrogenous pyrimidine base found in RNA but not in DNA.

Uridylate   A nucleotide in RNA in which the base is uridine.

V

Variable Numbers of Tandem Repeats (VNTRs) DNA sequence blocks of 2–60 base pairs which are repeated from two to more than 20 times in different individuals. This polymorphism makes VNTRs very useful DNA markers used in genomic mapping, linkage analysis and also DNA fingerprinting.

Variable expressivity Differences in the severity of expression of a particular genotype.

Variable region The portion of an immunoglobulin molecule that varies greatly in amino acid sequence among antibodies in the same subclass. It is also called as V region.

Variance A measure of the spread of a statistical distribution; the mean of the squares of the deviations from the mean.

variation (genetic) Variation in genetic sequences and the detection of DNA sequence variants genome-wide allow studies relating the distribution of sequence variation to a population history. This in turn allows one to determine the density of SNPs or other markers needed for gene mapping studies. Quantitation of these variations together with analytical tools for studying sequence variation also relate genetic variations to phenotype.

Variegation Mottled or mosaic expression.

vector Any agent that transfers material (typically DNA) from one host to another. Typically, DNA vectors are autonomous DNA elements (such as plasmids) that can be manipulated and integrated into a host's DNA or recombinant viruses.

Viability The probability of survival to reproductive age.

Viral oncogene A class of genes found in certain viruses that predispose to cancer. Viral oncogenes are the viral counterparts of cellular oncogenes.

virtual libraries The creation and storage of vast collections of molecular structures in an electronic database. These databases may be queried for subsets that exhibit specific physico-chemical features, or may be "virtually screened" for their ability to bind a drug target. This process may be performed prior to the synthesis and testing of the molecules themselves.

Virulent pathogen Virulent pathogen or parasite that has the potential to do serious harm to its host.

Virulent phage A phage or virus capable only of a lytic cycle; contrasts with temperate phage.

Virus An infectious intracellular parasite able to reproduce only inside of the living cells.

visualization The process of representing abstract scientific data as images that can aid in understanding the meaning of the data.

Von Hippel-Lindau disease Hereditary disease marked by tumours in the retina, brain, other parts of the central nervous system, and various organs throughout the body. The disease is caused by an autosomal dominant mutation in chromosome 3.

V-type position effect A type of position effect in *Drosophila* that is characterized by discontinuous expression of one or more genes during development; usually results from chromosome breakage and rejoining such that euchromatic genes are repositioned in or near the centromeric heterochromatin.

W

Wall of pleural sac A thin layer of tissue covering the lungs and the wall of the chest cavity to protect and cushion the lungs. A small amount of fluid, that acts as a lubricant, allows the lungs to move smoothly in the chest cavity during breathing.

Warfarin A drug that prevents blood from clotting. It belongs to the family of drugs called anticoagulants (blood thinners).

Wasting State of general ill health characterized by malnutrition, weakness, and emaciation. It occurs during the course of a chronic disease.

Watson–Crick pairing Base pairing in DNA or RNA in which A pairs with T (or with U in RNA) and G pairs with C.

Watson strand The upper strand of duplex DNA when the strands are conventionally represented as parallel horizontal lines with the 5´ end of the top strand at the extreme left.

Webbed digits A union of two or moredigitsthatisnormalinmanybirds (as kingfishers) and in some lower mammals (as the kangaroos) and that occurs in humans often as a hereditary disorder marked by the joining or webbing of two or more fingers or toes.

weight matrix The density of binding sites in a gene or sequence can be used to derive a ratio of density for each element in a pattern of interest. The combined individual density ratios of all elements are then collectively used to build a scoring profile known as a weight matrix. This profile can be used to test the prediction of the identification of the selected pattern and the ability of the algorithm to discriminate them from non-pattern sequences.

western blot Technique in which specific antibodies are used to identify their antigens from a mixture of proteins. Typically, these protein mixtures are first separated by electrophoresis and then transferred onto nylon sheets by electrotransfer. Radio-labelled or enzyme-linked antibodies are incubated with the sheets and unbound antibodies are washed away allowing the position of the bound antibody to be revealed by autoradiography or colour which is formed upon addition of a substrate.

Wheezing Difficulty to breathe usually with a whistling sound.

White blood cell A type of blood cell that does not contain haemoglobin. White blood cells include lymphocytes, neutrophils, eosinophils, macrophages, and mast cells. These cells are produced by bone marrow and helps the body to fight with infection and other diseases.

White matter White nervous tissue constituting the conducting portion of the brain and spinal cord.

White of eye The dense, fibrous, opaque, white-outer coat enclosing the eyeball except the part covered by the cornea called also sclerotic coat.

wild type Form of a gene or allele that is considered the "standard" or most common.

Wild-type allele The normal, as opposed to the mutant, gene or allele.

Wilms tumour A kidney cancer (tumour) that occurs in children, usually before age five.

Wobble The acceptable pairing of several possible bases in an anticodon with the base present in the third position of a codon.

Writers cramp A painful spasmodic cramp of muscles of the hand or fingers brought on by excessive writing called also graphospasm.

X

Xanthogranuloma A condition of faulty lipid metabolism in which yellow nodules of lipoid matter are deposited in the skin and mucosae, giving rise to granulomatous reactions.

Xanthoma A fatty irregular yellow patch or nodule containing lipid-filled foam cells that occurs on the skin (as of the eyelids, neck, or back) or in internal tissue and is associated especially with disturbances of lipid metabolism.

X chromosome In mammals, the sex chromosome that is found in two copies in the homogametic sex (female in humans) and one copy in the heterogametic sex (male in humans).

X-chromosome inactivation In females, the phenomenon by which one X chromosome (either maternally or paternally derived) is randomly inactivated in early embryonic cells, with fixed inactivation in all descendant cells. It was first described by the geneticist, Mary Lyon.

X-chromosome inactivation study Molecular genetic testing to assess the relative proportion of methylated (inactive) X chromosomes to unmethylated (active) X chromosomes. This study is used to determine if X-chromosome inactivation is random or skewed.

Xeroderma pigmentosum An inherited defect in the repair of damage to DNA, caused by UV light associated with extreme sensitivity to sunlight and multiple skin cancers.

X-linked dominant A dominant trait or disorder caused by a mutation in a gene on the X chromosome. The phenotype is expressed in heterozygous females as well as in hemizygous males (having only one X chromosome); affected males tend to have a more severe phenotype than affected females.

X-linked inheritance The pattern of hereditary transmission of genes located in the X chromosome; usually evident from the production of non-identical classes of progeny from reciprocal crosses.

X-linked lethal A disorder caused by a dominant mutation in a gene on the X chromosome that is observed almost exclusively in females because it is almost always lethal in males who inherit the gene mutation.

X-linked recessive A mode of inheritance in which a mutation in a gene on the X chromosome causes the phenotype to be expressed in males who are hemizygous for the gene mutation (i.e., they have only one X chromosome) and in females who are homozygous for the gene mutation (i.e., they have a copy of the gene mutation on each of their two X chromosomes). Carrier females who have only one copy of the mutation do not usually express the phenotype, although differences in X-chromosome inactivation can lead to varying degrees of clinical expression in carrier females.

X-ray computed tomography (CT scan) A series of detailed pictures of areas inside the body taken from different angles; the pictures are created by a computer linked to an X-ray machine. Also called computerized tomography and computerized axial tomography (CAT) scan.

X-rays A type of high-energy radiation. In low doses, X-rays are used to diagnose diseases by making pictures of the inside of the body. In high doses, X-rays are used to treat cancer.

X-ray therapy The use of high-energy radiation from X-rays, gamma rays, neutrons, and other sources to kill cancer cells and shrink tumours. Radiation may come from a machine outside the body (external-beam radiation therapy), or from materials called radioisotopes. Radioisotopes produce radiation and can be placed in or near the tumour or cancer cells. This type of radiation treatment is called internal radiation therapy, implant radiation, interstitial radiation, or brachytherapy. Systemic radiation therapy uses a radioactive substance, such as a radio-labelled monoclonal antibody, that circulates throughout the body. Also called radiotherapy, irradiation, and X-ray therapy.

Y

Yap 1p A yeast transcription factor encoded by the *yap 1* gene.

Y chromosome The sex chromosome present only in the heterogametic sex. In mammals, it is the male-determining sex chromosome.

Yeast Artifical Chromosome (YAC) An artifical chromosome containing a yeast centromere, two telomeric sequences, and a marker. The YAC is constructed by cloning very large genomic fragments (up to one million bases) from another species into yeast vectors that can replicate in yeast.

yeast 2-hybrid system A yeast-based method used to simultaneously identify, and clone the gene for, proteins interacting with a known protein. The basis of this method is a "transcriptional reporter assay" in which reporter gene expression is dependent on two domains. The first domain is linked to the known protein. The second domain is genetically linked to a library. If the library is screened against the known protein the two domains will interact only if a protein from the library binds the known protein, resulting in transcription activation of the reporter gene, and a blue colour. The "blue yeast clone" will contain the gene encoding the newly identified protein.

Yellow spot A small yellowish area lying slightly lateral to the centre of the retina—that constitutes the region of maximum visual acuity and is made up almost wholly of retinal cones.

Yoctomoles 1 ymol = 10^{-24} moles = 0.6 molecules. A quantity recently coined due to increased sensitivity in proteomics.

Z

Z-DNA A conformation of DNA existing as a left-handed double helix (the phosphate–sugar backbone forms a left-handed zigzag course), which may play a role in gene regulation.

Zeptomoles A quantity recently coined due to increased sensitivity in proteomics. 1 zmol $= 10^{-21}$ mol $= 602$ molecules.

Zinc finger domain Motifs in DNA- and RNA-binding proteins whose amino acids are folded into a single structural unit around a zinc atom. In the classic zinc finger, one zinc atom is bound to two cysteines and two histidines. In between the cysteines and histidines are 12 residues which form a DNA- binding fingertip. By variations in the composition of the sequences in the fingertip and the number and spacing of tandem repeats of the motif, zinc fingers can form a large number of different sequence-specific binding sites.

zinc fingers A protein motif formed by the interaction of repeated cysteine and histidine residues with a zinc ion. The spacing of the repeats results in finger-like arrangements of the protein loops formed from the interaction which interact with DNA. These motifs are typically found in transcription factors.

Zygosity testing The process through which DNA sequences are compared to assess whether individuals born from a multiple gestation (twins, triplets, etc.) are monozygotic (identical) or dizygotic (fraternal); often used to identify a suitable donor for organ transplantation or to estimate disease susceptibility risk if one sibling is affected.

zygote A diploid cell created by the union of a male and female gamete.

Zygotene The substage of meiotic prophase I in which homologous chromosomes synapse.

Zygotic gene Any group of genes that control early development through their expression in the zygote.

www.ingramcontent.com/pod-product-compliance
Lightning Source LLC
LaVergne TN
LVHW012043200726

843506LV00021B/743